CONTEMPORARY SOCIAL

RESEARCH METHODS

A Text Using MicroCase®

Rodney Stark

UNIVERSITY OF WASHINGTON

Lynne Roberts

THIRD EDITION

WADSWORTH
THOMSON LEARNING™

Australia • Canada • Mexico • Singapore • Spain • United Kingdom • United States

Sociology Editor: *Lin Marshall*
Assistant Editor: *Analie Barnett*
Editorial Assistant: *Reilly O'Neal*
Technology Project Manager: *Dee Dee Zobian*

Marketing Manager: *Matthew Wright*
Production Service: *Jodi Gleason*
Copy Editor: *Margaret Moore*
Printer: *Transcontinental Printing, Inc.*

For more information, contact
Wadsworth/Thomson Learning
10 Davis Drive
Belmont, CA 94002-3098
USA

For more information about our products, contact us:
Thomson Learning Academic Resource Center
1-800-423-0563
http://www.wadsworth.com

International Headquarters
Thomson Learning
International Division
290 Harbor Drive, 2nd Floor
Stamford, CT 06902-7477
USA

UK/Europe/Middle East/South Africa
Thomson Learning
Berkshire House
168-173 High Holborn
London WC1V 7AA
United Kingdom

Asia
Thomson Learning
60 Albert Complex, #15-01
Singapore 189969

Canada
Nelson Thomson Learning
1120 Birchmount Road
Toronto, Ontario M1K 5G4
Canada

ISBN-13: 978-0-534-58189-3
ISBN-10: 0-534-58189-7

Contents

About the Authors

Rodney Stark began his career as a newspaper reporter and then received his Ph.D. from the University of California, Berkeley. While in Berkeley, he spent eight years as a Research Sociologist at the Survey Research Center and then joined the staff of the Center for the Study of Law and Society where, for two years, he directed and conducted field research on student protest and on the police. He left Berkeley to become Professor of Sociology at the University of Washington in Seattle where, for nearly a decade, he taught the required two-term, graduate social research methods course. Stark is the author or coauthor of 21 books and more than 130 scholarly articles.

Lynne Roberts received her Ph.D. from Stanford University where she also was on the staff of the Computing Center. She left Stanford to inaugurate a year long research methods course in the then-new doctoral program offered by the School of Social Welfare at the University of California, Berkeley. After several years, she joined the faculty of the University of Washington in Seattle where she served as Associate Professor of Sociology. Roberts left academic life to found MicroCase Corporation, a software publishing company. During her tenure at Washington, Roberts taught statistics at both the undergraduate and graduate levels. Trained as an experimental social psychologist, she designed and conducted many experiments, the results of which she reported in a book and a number of scholarly articles. She also has published articles on methodological subjects including scaling and experimental design.

Preface: To the Instructor

We are delighted with the response to the first two editions of *Contemporary Social Research Methods*. Our desire was to improve the way social science methods are taught and, indeed, many faculty have told us that this book has forever changed their research methods course.

We developed this package because we found it impossible to show students how research actually is done from books that ignore analysis, or merely tack-on several analysis chapters at the end. It's as if one could teach woodshop by showing students some hammers, saws, chisels, and sanders, without ever using them to *do* anything. How can one deal with issues concerning various aspects of data collection when students don't know what the data are for? So, we wrote a textbook and laboratory package in which basic analysis is at the front—in the workbook students become familiar with real data analysis so that by the time they encounter various forms of data collection, they will be guided by a sense of how data can and should be used.

Moreover, in this course data will be used! As we tell students in the preface addressed to them: *welcome to the real world of social research*. We have made it possible to shift the methods course away from passive learning, to active doing. And by now hundreds of instructors have discovered that it works—that students do learn much more (and much more eagerly) when they are able to explore real issues with real data.

When we began the first edition, our aim was to create a book that would work for us—each of us has taught methods at both the graduate and undergraduate levels. We also felt qualified to do the book because each of us had done a substantial amount of research using each of the major research methods. But we were neither so arrogant nor reckless not to have had every statement concerning the proper use, interpretation, or limits to statistics read carefully and approved by prominent statisticians. In similar fashion, the book has been screened by experts in sampling, questionnaire construction, time series analysis, field research, and many other topics covered in the book, including theory. If you discover things in the book that are contrary to what you were taught, welcome to the club—we learned a great deal while writing the book. The final result is a package of materials that not only is contemporary, but one we believe to be authoritative.

Reviewers and Consultants:

William Sims Bainbridge
Sociology, National Science Foundation

Lawrence F. Jones
Government, Angelo State University

Steven Barkan
Sociology, University of Maine

Eileen Barker
Sociology, London School of Economics

Steven Bird
Sociology, Taylor University

Allen B. Brierly
Political Science, University of Northern Iowa

Sue Crull
Family & Consumer Sciences, Iowa State
University

Roger Finke
Sociology, Purdue University

Norval D. Glenn
Sociology, University of Texas at Austin

Frances L. Hoffmann
Women & Gender Studies, University of
Missouri-St. Louis

Laurence R. Iannaccone
Economics, Santa Clara University

Patrick D. Nolan
Sociology, University of South Carolina

David Nordlie
Sociology, Bemidji State University

Robert M. Ogles
Communication, Purdue University

Michelle Piskulich
Political Science, Oakland University

Adrian Raftery
Statistics & Sociology, University
of Washington

Darren Sherkat
Sociology, Vanderbilt University

Peter Sinden
Sociology, SUNY College at Fredonia

Tom W. Smith
Co-director of the General Social
Survey, NORC

In addition to all of the expert advice we received from colleagues before we published the first edition, we have received very useful comments from some of those who adopted it.

Consultants for the Second Edition:

Joni Boye-Beaman, Wayne State College; Tony Cappen, University of Pittsburgh; Judi Caron-Sheppard, Norfolk State University; Susan Collins, Iowa State University; Sue Crull, Iowa State University; David Decker, California State University-San Bernadino; Michael Donahue, Armstrong State College; Clifton Flynn, University of South Carolina-Spartanburg; Bob Hall, West Virginia State College; Azi Jahanbegloo, Capital University; Robert Miller, University of North Carolina-Wilmington; Roger Nemeth, Hope College; Alden Roberts, Texas Technical University; Kay Banister Schaffer, St. Cloud State University; Brent Shea, Sweet Briar College; Carl Simpson, Western Washington University; Edward J. Steffes, Salisbury State University; Jim Wiest, Hastings College; Richard Zeller, Bowling Green State University.

Consultants for the Third Edition:

Sue R. Crull, Iowa State University; Barbara R. Keating, Minnesota State University, Mankato; James G. Leibert, Dickinson State University; Bonnie L. Ross, Chapman University; Roy C. Treadway, Illinois State University; Douglas M. Wiig, Grand View College.

Preface: To the Student

Welcome to the real world of social science research. In this course you will encounter no make-believe and very few hypothetical examples. You are not only going to read about research, you are going to *do it*. In addition to offering a basic textbook that covers contemporary social science methods, this instructional package includes a workbook of exercises (many of them requiring a computer), a student version of the MicroCase Analysis System, and several major data files. All the data are real. In fact they are some of the best data available to professional researchers, and you will learn to use the same techniques they use.

The software is so easy to use that you will learn it without study—so you can devote your study time to more important things. The workbook assignments will lead you step-by-step through each feature of the software as it becomes relevant. So, just follow along and soon you will be able to do anything you want to do (and you also will know why you want to do it).

Despite being easy to learn, this software is not a toy. Its computational heart is the same as that included in the full MicroCase Analysis System. However, because the full system has many capacities and features that you won't need, only the essential functions are included in the student version.

Instructors are always telling students they will get more out of a course if they read a chapter before it is the subject of a lecture. Nevertheless, many students often wait until the week of an exam to do their reading. That really won't work in this course. Although the workbook exercises review basic principles and concepts from the text, it will be difficult to understand many of the exercises if you have not read the chapter on which they are based.

When we created this package, we knew that most students who will use it will not go on to be professional researchers. With that in mind, we have tried to maximize the general educational value of the course—to let you take away some worthwhile knowledge. This goal will have been achieved if the course prepares you to be an informed consumer, able to recognize at a glance the unreliable and silly claims among the flood of social science results reported in the media.

Should you care to send us comments, suggestions or (especially) complaints, write us at:

Wadsworth Group/Thomson Learning
10 Davis Drive
Belmont, CA 94002-3098

Concepts and Theories

It seems as though the news media report a hideous murder case almost every day. An estranged husband guns down his wife and three children in front of a dozen witnesses. A women is convicted of murdering her five children, one-by-one, over a period of years. A postal employee takes an assault gun to work and kills his boss and two fellow workers.

As they bring us these stories, the media raise many questions. Why was the estranged husband still on the loose despite many previous threats against his family? Why did it take so long before medical examiners realized that this woman was smothering child after child? Where did the assault gun come from?

These are the questions of journalists, not social scientists. Granted that most social scientists personally find these questions of interest, but professionally, no individual case is very important to them. Social scientists are concerned with far more general questions prompted by such things as their awareness that the cases receiving media attention are but a tiny fraction of the more than 15,000 homicides that occur in the United States every year. Thus, social scientists don't ask why a particular person was murdered but rather, why is the homicide rate so much higher in the United States than in Canada? Is the murder rate falling? If so, why? And why is it that so many murder victims are killed by relatives or friends? Why do people abuse their spouses? Do people of different racial and ethnic backgrounds hold different views about what is a crime? Does race influence attitudes toward the criminal justice system—the police, courts, and prisons?

If you compare the examples of questions journalists ask with those social scientists ask, it will be obvious that, while journalists *particularize*, social scientists *generalize*. Journalists want to know who the murder victims were and who has been brought in for questioning. Social scientists want to know why murder rates in general are higher in some times and places than in others, or why some people are much more apt than others to be murdered.

Like journalists, however, social scientists not only ask questions, they also try to discover answers to their questions. In this course, you are going to learn how social scientists seek their answers. For example, how can we discover why some people are prejudiced, why some vote for liberal candidates, or why some people watch so much TV? How can we know if a college education really pays off for most graduates or if unemployment is an important cause of divorce?

The process of seeking reliable answers to questions such as these is called *research*. Social scientific research involves applying a few basic principles and techniques social scientists have developed to help them discover how things really are and why they are that way. The purpose of this course is to explain these principles and techniques—to introduce you to social science research methods. And the purpose of this chapter is to introduce you to the fundamental building blocks of social science: concepts and theories, indicators, and hypotheses.

Doing Social Science

Despite the fact that you are reading these lines in a conventional-looking textbook, this is not going to be a "book learning" course. Instead, this is going to be a hands-on course in which you learn by doing, by actually using the major social research methods to examine data of the highest quality.

This textbook, the accompanying workbook, the student version of the MicroCase Analysis System, and databases will help you learn how to do social science by doing it yourself, using real data to answer real questions. Better yet, MicroCase is extremely easy to use. So, even if you have never used a computer before, don't worry. Thousands of students before you have had no trouble using MicroCase, and the instructions in the workbook will show you step-by-step exactly what you need to do. However, despite being easy, MicroCase is not a toy. It is a very powerful research tool that allows you to use the same techniques employed by professional social researchers.

Studying Ourselves

A major difficulty facing social scientists as opposed to natural scientists can be summed up this way: Molecules don't tell lies or get grumpy, but people do. Chemists needn't worry that their chemicals might one day decide to fake their reactions or refuse to participate in an experiment. But, when self-conscious beings are the primary object of study, the accuracy of observations becomes an issue. Are the people being observed behaving in a normal fashion? Are people giving truthful answers or providing honest statistics? A substantial number of social science research techniques are devoted to limiting the likelihood of being fooled.

A second difficulty involves observer bias. Social scientists must try to make sure that what they see isn't simply what they want to see. Biologists probably have no preferences about the color of a salamander's bile ducts, but social scientists tend to have strong opinions about human nature. Thus, they must use methods that minimize the possibility that they are seeing things as they would like them to be, rather than the way they truly are.

However, there is an advantage to the fact that social scientists are, in effect, studying themselves. Everyone knows a lot about people without having to learn it in school—if we couldn't accurately predict how other people will respond in common situations, social life would be impossible. We all know that most people would rather be greeted with a smile than with an obscene gesture, and that when we go to the bank the tellers will cash our checks or accept our deposits and not throw fruit at us or paint our automobiles. Aspiring social scientists begin with a substantial understanding of their subject. For this reason, much that turns up in social scientific research was obvious before the research was done. Consider the following examples:

- Burglars more often break into the homes of middle- and upper-income people than into the homes of the poor.
- College-educated people are substantially less likely to attend church weekly than are persons who did not complete high school.
- Older people are less likely than younger people to read the newspaper daily.
- African Americans are far more likely to drink alcoholic beverages than are whites.

Whether or not these statements struck you as obvious, the results of many national surveys all show that *each of them is wrong!* Burglars are far more active in the poorest neighborhoods. The more education people have, the more likely they are to attend church weekly. Older people are far more likely than younger people to read a newspaper every day, and African Americans are much less likely to drink than are whites.

This is one reason social scientific research is valuable: Too often, what many regard as obvious turns out to be obviously wrong.

DESCRIPTION AND EXPLANATION

No matter what social scientists study or how they study it, their work is guided by two primary goals: description and explanation. That is, social scientists attempt to accurately *describe* some aspect of the world—in what parts of the United States and Canada occult and psychic activities are more common, for example, or what percentage of people in a particular population support capital punishment. Social scientists also seek to *explain* why things are the way they are—why do occult activities thrive in some places but not in others? Why are men more apt to support capital punishment than are women?

We cannot describe, let alone explain, everything going on in the world at once. Thus, all scientists, whether social, physical, or natural, take things apart and group them according to some principle or characteristic. Chemists divide things into elements, biologists often classify life forms according to species, and social scientists may classify people on the basis of characteristics such as social class, age, or gender. Scientists use *concepts* to identify and describe parts of reality with which they are concerned. They use *theories* to put the pieces back together and to explain how things work.

CONCEPTS

Scientific **concepts** are abstract terms that identify a class of "things" to be regarded as alike. Consider the concept of *mammal*, defined as a warm-blooded animal that gives birth to living young (as opposed to laying eggs). This is an **abstraction** in that we can't see the *concept* of mammal. It is an intellectual creation, a definition, existing only in our minds. All we can see are *instances* of this concept: actual animals that belong to this class such as dogs, cats, mice, whales, horses, and sheep. Because scientific concepts are abstract, they apply to all possible members of the class, all that have been, are, shall be, or could be—extinct mammals as well as those not yet in existence.

Notice that there is immense variation among the members of the class of things identified by this concept: in size, temperament, and even eating habits. But for the purposes of biologists, these differences are ignored in favor of the similarities among the members.

As fundamental building blocks of science, concepts must be clear and efficient. The following principles guide the construction of good concepts.

Parsimony. Not only are concepts abstractions, but concepts vary in their degree of abstraction. For use in theories it generally is the case that the *more* abstract a concept is, the better. The concepts of mammal and of dog are both abstract. But mammal is more abstract in the sense that it applies to a much larger set of "things that are alike." If biologists can formulate a theory that applies to mammals, their theory explains more than if it applied only to dogs, and thus separate theories are needed for cats, rats, elephants, and donkeys. In similar fashion, a theory that utilizes the concept of prejudice will explain more than one restricted to the less general concept of anti-Semitism.

This rule merely reflects the commitment of all sciences to **the law of parsimony**, or what is sometimes referred to as **Ockham's razor**. William of Ockham (circa 1285–1349), an English philosopher, taught that the elements of scientific explanations "are not to be multiplied beyond necessity." This principle is called his "razor" because he applied it to cut away large amounts of "unnecessary" elements from philosophical arguments popular in his time.

As used by modern philosophers of science, the law of parsimony reads as follows: *Theories always should attempt to explain the most with the least.* We should try to explain as much as possible with a theory that is as simple as possible. Applied to concepts, parsimony encourages greater abstraction.

Utility. When we wish to know the definition of a word, we often look it up in the dictionary. Because dictionaries are regarded as authoritative, we easily are encouraged to believe that we can distinguish between true and false definitions— that the dictionary provides the "real" meaning of the word. But this isn't so. All definitions are merely conventions, and the definitions of words often change over time. As Carl G. Hempel (1952) pointed out in his classic work on scientific concepts, it sometimes is claimed that "real" definitions somehow are an intrinsic part of the thing being defined, but this is folklore. Scientific definitions are not thought of as real in this sense at all, but are regarded as names that simply are *assigned* to something. To underscore this point, Hempel identified all scientific concepts as **nominal definitions**—that they are merely names.

The ultimate test of all concepts is in their **utility**, or their usefulness in constructing efficient theories. *What works better, is better!* Recognize that the search for good concepts cannot be guided by truth, for nominal definitions are neither true nor false, but some are far more useful than others.

Since concepts are mental abstractions, we are free to construct them as we please. To say that the concept of prejudice refers to holding negative beliefs about people because of their race or ethnicity is simply a definition. Definitions are based on convention, not on truth. Of course, it is true that this is how the concept of prejudice is being defined, but there is no basis for saying that this is the true definition of prejudice. Indeed, we might wish to change this definition to include the qualification that the negative beliefs must be false, or that they must be accompanied by hostile feelings. None of these alternative definitions is truer than the others. And, since they are equally clear, they can be compared only on the basis of how well they work in a theory.

Even concepts that lack any application to reality usually are not really false, although they may be useless. Consider the unicorn. Let's define the concept of *unicorn* as applying to all horselike animals having a single horn growing from the middle of their foreheads. So far as we know, no unicorns ever have existed. That doesn't make the concept false, however, and if we ever do discover a unicorn we already will know what to call it.[1] But, until such a time, the concept lacks utility. However, if we added to the definition of the concept that unicorns *are known to inhabit the veldt country of South Africa*, then the concept would be false in addition to being useless.

Clear Boundaries.   Efficient concepts will have clear boundaries, delineations that eliminate ambiguity about what a concept does and does not include. The concept of *suicide* posed boundary problems for nineteenth-century social scientists, including Emile Durkheim (1897). Durkheim noted that the word *suicide* was used very imprecisely in everyday speech. In addition to applying to people who took poison or jumped off bridges, the word often was applied to actions that resulted in death but that were not done in pursuit of death—as in the case of a "suicidal" military attack. Durkheim was certain that this was too inclusive a concept, mixing actions having quite different causes. He wanted to clearly mark the boundaries of the concept. Consequently, Durkheim began his study by limiting his concept of suicide to only those cases where the primary aim of the person committing the act was to end his or her own life.

Naming Is Not Explaining.   Concepts are extremely useful for classification, but it is important to realize that concepts do not explain anything. Simply to name things does not tell us why they exist, what they do, or how they work. The concept of *mammal* does not explain *why* some creatures give birth to living young, nor does the concept of *anti-Semitism* explain why some people hate Jews. To pursue this issue further, consider a concept often used to identify certain religious groups. *Sects* are religious groups in a relatively high state of tension with their social environments (Johnson, 1963). That is, sects are religious groups that impose

---

[1] The first European explorer to see a rhinoceros concluded that here at last was the unicorn. He acknowledged that while the ancient stories had been correct about a creature with one horn, they had failed to convey how ugly the critter was.

significantly stricter moral codes on their members than does the surrounding society. For example, many Protestant sects prohibit members from drinking, gambling, engaging in premarital sex, or having abortions. But the concept of sect does not tell us *why* they do so. Nothing is explained by the statement that a group imposes stricter moral rules *because* it is a sect, since that is the definition of a sect—that's like saying a sect is a sect because it is a sect.

The concept of sect does nothing more (nor less) than allow us to classify various religious bodies. Explanations involving this concept do not reside in the concept itself. For example, it is well known that sects, especially if they are successful, tend to move from a higher to a lower state of tension—sects often are transformed into churches. But no explanation of this transformation can be found in the definition of sects.

Attempts to substitute names for explanations are what logicians refer to as *tautologies*. A **tautology** is any statement that is necessarily true by definition. A classic example is President Calvin Coolidge's famous remark that "as more and more men are thrown out of work, unemployment results." Another name for a tautology is a circular argument, one that turns back on itself—for example, "Why does it rain? Because of the weather."

Many tautologies are easy to recognize. But some tautologies are less obvious because they lurk beneath the surface of concepts having different names but identical or nearly identical definitions. In an effort to explain why people smoke marijuana, Howard S. Becker (1953:241) claimed,

> *A person, then, cannot begin to use marijuana for pleasure, or continue its use*
> *for pleasure, unless he learns to define its effects as enjoyable, unless it becomes*
> *and remains an object which he conceives as capable of producing pleasure.*

The apparent profundity of the statement vanishes when we realize that the two key terms *pleasure* and *enjoyable* are synonyms and that, reduced to basics, the sentence would read, "Pleasure must be pleasurable."

Or consider the "explanation" by an anthropologist that the Chinese do not like milk because of their culture, in light of his definition of culture as the "way of life" of any given group or society (Kluckhohn, 1949). Since not liking milk is part of the Chinese way of life, here too is a tautology posing as an explanation. Our understanding of *why* the Chinese do not like milk is not increased by invoking the concept of culture. As George Homans (1967:12–13) noted,

> *What we should have liked to know was why milk, specifically, rather than,*
> *say, tea was disliked. Talking about culture did not answer this question at*
> *all—not at all. More generally, "explanation by concept" is not explanation.*

If the goal is not merely to classify some set of phenomena, but to *explain* them, then more is required than a concept or even a large set of concepts—for these add up to nothing more than a set of parts. To explain, it is necessary to say why and how the parts or concepts fit together and function. The difference here is between a parts catalogue and a working diagram of an engine. Working diagrams involving scientific concepts are theories.

THEORIES

Few words are more misused and misunderstood than the word *theory*. Often it is used as a synonym for "groundless speculation," as in "But that's merely a theory." This assumes that *important* and *abstract* statements can be made about the world that are most certainly true as opposed to similar statements that remain to be proven. Abstract statements that are certainly true also are certainly unimportant—for such certainty to exist, the statements must be tautologies. Conversely, as we shall see, abstract statements that are not tautologies can never finally be proven. Whether it is the theory of gravity, the theory of evolution, or the demographic transition theory, each remains "only a theory" in the sense that we never can be absolutely certain that it is true (but we can become very confident).

Another abuse of the word *theory* is to confuse it with *conjecture* as in "My theory about why Becky and John broke up is that . . . ." This is merely the statement of an opinion about a specific event. Such a statement can, at least in principle, be fully verified, but it has no application beyond the single instance.

As used in science, the word *theory* refers to a particular kind of statement designed to explain something of general interest and application. These statements have two essential features: They are *abstract*, and they are *falsifiable* in that it is possible to say what evidence would show them to be false.

Abstract. **Theories** are abstract statements that say *why* and *how* some set of concepts are linked. Their purpose is to *explain* some portion of reality.

Let's consider a recent example from criminology. The phenomenon to be explained is that some neighborhoods have high crime rates decade after decade despite the fact that there are frequent turnovers in their populations. For example, there are urban neighborhoods, especially in eastern cities, that had high crime rates way back when they were Irish neighborhoods, and the crime rates did not decline when they became Italian neighborhoods, nor when they became Polish neighborhoods, nor when they subsequently became African-American neighborhoods. The *theory of deviant places* attempts to explain this persistence of high crime rates on the basis of aspects of the neighborhood and how people respond to these aspects rather than on the basis of the "kinds of people" who live in these neighborhoods (Stark, 1987).

The entire theory consists of 30 propositions including several dozen primary concepts. Here it is sufficient to consider only a few of each.

> **Proposition 1:** The greater the density of a neighborhood, the more association between those most and least predisposed to commit crimes.

Concepts:
Density: the number of residents per some measure of area.
Neighborhood: an area defined on the basis of some measure of proximity.
Association: interpersonal contact.
Crimes: acts that violate the law.

Amplification: The denser the neighborhood, the harder it is for people to avoid one another and thus the more likely it is that "good" people will be in constant contact with the "bad" people.

Proposition 2: The greater the density of an area, the higher the level of moral cynicism.

Concept:
Moral cynicism: the belief that people are much worse than they pretend to be.

Amplification: The denser the neighborhood, the harder it will be for people to shield their private lives. In dense neighborhoods, the neighbors overhear family fights, for example, whereas in spacious suburbs, they do not.

Proposition 17: It is socially stigmatizing to live in a dilapidated neighborhood.

Concepts:
Social stigma: a trait drawing social disapproval or suspicion upon the individual.
Dilapidation: the state of being worn out, run down, decayed, trashy

Amplification: If you live in a dilapidated neighborhood, people will think less of you.

Proposition 21: Those residents who are more successful and who could serve as the best role models will flee stigmatized neighborhoods.

Concept:
Role model: someone whose behavior sets a standard for others.

Amplification: Those who can escape bad neighborhoods do.

Notice that the concepts in these propositions are abstractions. No one can see or touch the concept of *role model*, although we can see many concrete instances of this concept. Nor can we see the concept of *density*, although we can easily experience variations in actual degrees of density. Because theories contain concepts and concepts are abstractions, all scientific theories also are abstractions. That is, you can't see or touch a theory. We can visit dilapidated neighborhoods and try to discover whether people there are unusual in their degree of moral cynicism, but the concepts of dilapidation and moral cynicism are entirely in our minds. Consider that if we changed the definition of *dilapidation* to mean "beautiful," all neighborhoods would continue to look precisely the way they did before the redefinition. All that would change is that we would call different neighborhoods dilapidated than we did before.

Not all social scientific theories are stated in terms of formal propositions (although they can be translated into propositions). The *demographic transition theory* offers a useful example. Throughout most of human history, population size was determined by high levels of fertility offset by high levels of mortality. We can translate that as "People had a lot of kids and life expectancy was short, especially due to high rates of infant and child deaths." Then, in response to industrialization, mortality rates began to fall in Europe and North America. At first, this caused rapid population growth. Soon after, however, population growth was halted by a decline in fertility. In the industrialized nations today, population size is determined by low fertility, offset by low mortality.

This shift, or transition, from high fertility and mortality to an era having low fertility and mortality is known as the *demographic transition*. Modern demographers have constructed a theory to explain why and how this transition occurred (Davis, 1945; Berelson, 1978; Cutright and Hargens, 1984). As first stated by **Kingsley Davis** (1945), the theory begins with the assumption that people seek to maximize—that we try to select the choice that will yield us the greatest benefit at the least cost. In preindustrial situations, the rational family will maximize fertility. They will do so for two primary reasons. First, high mortality rates require them to "stockpile" children in the anticipation that many will not survive to adulthood. Second, in agricultural societies, children are an economic benefit as a source of cheap labor when they are young and later as a source of support/for elderly family members. Industrialization changed the equation. Low mortality removed the need to stockpile children. Meanwhile, children became an economic burden to urban families, costing far more to raise than any possible economic benefits they could provide. Finally, primary responsibility for the support of the elderly was transferred from children to pension plans and government benefits.

This theory generates many empirical predictions, and testing these predictions has occupied many demographers for many years as they have monitored rates of industrial development and checked to see if the predicted shifts in fertility occurred. So far, so good.

Falsifiable.   In addition to being abstract statements that attempt to explain why and how sets of concepts fit together, to qualify as a theory, the set of statements must be **falsifiable**—that is, a real theory directs our attention to observations that would prove the theory to be false; it implies empirical predictions and prohibitions.

Empirical means "observable through the senses." While it is true that we can't see or touch theories, that they exist only in our minds, it also is true that a set of statements linking a set of concepts qualifies as a theory *only* if it tells us that certain observable things must or must not happen. That is, a theory will tell us what to look at, when to look, and what to expect to see. We test theories by checking up on these observable predictions. When the observations disagree with the predictions made from the theory, we know that the theory is incorrect—that we have falsified it in whole or part.

Consider the demographic transition theory. If a number of industrializing nations achieved low levels of mortality and their fertility rates did not soon begin to decline, we would know that the theory was wrong because a predicted outcome did not occur and a prohibited outcome did. Or, in the case of the theory of deviant places, if moral cynicism turned out to be *lower* in denser neighborhoods than in less dense neighborhoods, that part of the theory would be falsified.

The immense benefit of requiring that, in principle, theories be falsifiable is that it permits us to eliminate our mistakes. When it becomes clear that the empirical predictions and prohibitions derived from a theory are incompatible with the appropriate observations, we know it is time either to repair the theory or to scrap it entirely and try again. What we must not do is to modify the theory so that it becomes compatible with *all possible* empirical observations.

In psychology there once was considerable interest in what was called *frustration/aggression theory*. This theory proposed that, when people become sufficiently frustrated, they respond by becoming aggressive—they vent their frustrations as aggressive behavior toward the sources of their frustration. All was well until repeated experiments in which people were placed in frustrating circumstances failed to find the predicted aggressive responses. In response, the theory was modified by the introduction of an additional concept: *sublimation*. This concept identified aggressive responses to frustration as subject to concealment. That is, people did not always respond aggressively to frustration. Instead, their aggressive impulses could be short-circuited as they were sublimated or transformed into socially acceptable responses. This solution seemed to satisfy many supporters of the frustration/aggression theory. In fact, however, this addition changed frustration/aggression from a theory into a tautology. No longer was anything predicted or prohibited. Following frustration, we should either expect to observe aggression or expect to not observe it. That "prediction" must always come true.

A primary reason for doing research is to test or empirically check out theories. But, if research results can falsify a theory, *no* amount of research ever can prove that a theory is correct! As Rudolf Carnap (1953:48) explained, no "universal sentence" such as a "law of physics or biology" can ever be proven for "the number of instances to which the law refers . . . is infinite and therefore can never be exhausted." Theories, like concepts, apply not only to the past and present, but also to the future. Unless or until we run out of future, we never can exhaust the opportunities for a theory to be falsified. Consider the theory of gravity. It predicts that if we throw a ball up into the air, it will come down. The theory prohibits the ball from simply floating in space. So far, no one knows of an instance when the prediction and the prohibition failed. But it *could* happen tomorrow or in the next century or sometime or (probably) never. No one really expects the theory of gravity to fail, but the theory still has not been proved—although we plan our lives on the assumption that it is true.

The reason we place so much confidence in the theory of gravity is that it has so successfully resisted efforts to falsify it—after centuries of opportunities for it to fail, we have not yet observed a failure. The same rule applies to all theories. The more often and the more stringently they have been tested without failing, the more confidence we place in them (Carnap, 1953). Consequently, the proper aim for social scientific researchers is to do their best to falsify theories. To the extent that theories resist such efforts they gain credibility.

CONCEPTS AND INDICATORS

Concepts classify some observable portions of the world as being alike. When we say that theories must predict and prohibit certain observable things, what we are saying is that theories tell us to look at actual instances of various concepts and to expect certain relationships or connections among them. Suppose our theory included the prediction that the higher his or her social class, the more likely a person is to be a political conservative. *Social class* is a concept. So is *political conservative*. We can't observe concepts. But we can observe specific instances of concepts.

An **indicator** is an observable measure of a concept. Annual income would be a good indicator of a person's social class, so would his or her years of education or occupational prestige. Voting for particular political candidates, supporting certain political policies, or even subscribing to some magazines would be indicators of political conservatism.

THEORIES AND HYPOTHESES

Theories are abstract statements about reality. We test theories by checking on their observable predictions and prohibitions. We refer to these observable predictions as *hypotheses*.

A **hypothesis** tells us what to expect to observe when we examine relationships among indicators. That is, a theory specifies relationships among concepts, and the hypothesis derived from a theory specifies the relationships to be observed among indicators. In the example above, the theory specified a positive relationship between social class and conservative political views. We can't observe such a statement, it is an abstraction. But as with all theories, it tells us what to expect to see if we look at specific, observable indicators of its constituent concepts—in this case, indicators of social class and political conservatism. For example, historians examining support for royalist parties in nineteenth century Europe would, following this theory, expect to discover that wealthier people would have been more likely than factory workers to support royalists. Or we might expect to observe that today support for the British Labour Party is concentrated among the poor. Such observable expectations are hypotheses. Figure 1.1 illustrates the connection between theories and hypotheses.

Figure 1.1   Theory and Hypothesis

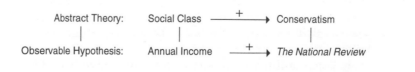

The theory predicts that there will be a positive relationship between indicators of social class and indicators of political conservatism—in this case between annual income and reading *The National Review* (a magazine of conservative opinion). Put another way, we are predicting a positive correlation between these two indicators.

CORRELATION

Correlation means "to go together, to vary in unison." Correlations can be either positive or negative.

The hypothesis above predicts a positive correlation. However, if it turned out that the higher their income, the *less* likely people are to read *The National Review*, the correlation would be negative. There are a number of ways to actually measure the degree to which indicators are correlated, and you will learn about these in the workbook exercises.

THE NULL HYPOTHESIS

When we test a hypothesis we risk making either of two errors, known as *type one* and *type two*. The **type one** error involves accepting that the predicted correlation exists, when in fact it does not. The **type two** error involves concluding that the predicted correlation does not exist, when in fact it does. Of the two types, scientists worry more about type one. As an additional guard against it, they follow a principle known as *testing the null hypothesis*. That is, hypotheses are always stated in the form denying the prediction derived from a theory. If a theory tells us to expect upper income Americans to be Republicans, for example, the null hypothesis would be: upper and lower income Americans are equally likely to be Republicans—that no relationship or correlation will be observed.

It is the null hypothesis that is the real focus of research testing. That is, to find that the data actually do support a theory it is necessary to *reject* the null hypothesis. The results must be sufficiently powerful so as to clearly overcome the likelihood of a type one error. We would rather be wrong by failing to confirm a hypothesis (by accepting the null hypothesis) than to be wrong by confirming a false hypothesis. In this sense, the burden of proof always is on the theory and close calls always go against the theory. Standards by which the null hypothesis is accepted or rejected will be discussed in Chapter 4, when we introduce the topic of statistical significance. Of course, a hypothesis also must be rejected if we find that the indicators are correlated, but in the *opposite way* predicted by the theory. Thus, we accept the null hypothesis (and thereby reject the hypothesis) if, for example, there is no correlation between social class and conservative political views, but we also reject the hypothesis if it turns out that lower-income people are more conservative than upper-income people.

Keep in mind too that even when we make an empirical test of a hypothesis derived from a theory and the results are as predicted, this does not *prove* the theory. The very next time we test a hypothesis derived from that same theory, the predicted results may not be found causing us to accept the null hypothesis.

Moreover, hypothesis testing does not end when we decide to accept or reject the null hypothesis, when we see whether or not an empirical prediction from the theory is confirmed. The diagram in Figure 1.1 only shows how hypotheses and indicators derive from a theory and its concepts; it shows only half of the relationship between theories and empirical observations. For the fact is that the results of research can cause us to modify theories, and sometimes unexpected research results even can prompt social scientists to formulate new theories.

Figure 1.2 illustrates the full relationship between theory and research. Theories provide hypotheses, which, in turn, are tested by observations. The results of these observations are summarized as an **empirical generalization**—a summary statement based on empirical observations. If the empirical generalization supports the hypothesis, our confidence in the theory is increased. If the generalization causes us to reject the hypothesis, the theory also is called into question. Finally, if repeated tests continue to force us to reject the hypothesis, the theory will be abandoned or modified.

Figure 1.2   The Wheel of Science

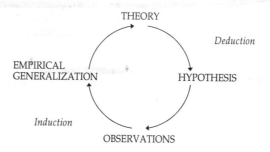

Adapted from Walter L. Wallace, *The Logic of Science in Sociology.* New York: Aldine Publishing Co. Copyright 1971 by Walter L. Wallace. Used by permission.

DEDUCTION AND INDUCTION

Notice that the words *deduction* and *induction* appear in Figure 1.2. These words refer to different modes of reasoning.

Deduction involves reasoning from the general to the specific—from a known (or assumed) principle or principles to an unknown but observable conclusion. Put another way, when scientists use deductive logic they begin with a general, abstract premise(s) or proposition(s) and then show that less general, empirical predictions are implied by the statement according to the rules of logic. Consider this example:

Major premise (or theoretical proposition): Conservatives oppose government ownership of industries.

Minor premise: Tom is a conservative.

Conclusion: Therefore, Tom opposes government ownership of industry.

By the rules of logic, if the major premise or theoretical proposition about conservatives is true, and if the minor premise which identifies Tom as a conservative is accurate, then the conclusion *must be true*. This has a very important implication.

When hypotheses are deduced from theories, as illustrated in Figure 1.2, if the theory is true (even though we can never know for sure that it is), then the hypothesis must be confirmed by observations. And that is precisely the reason that a failure of the observations to confirm a hypothesis has such an important implication for theories. For, if the hypothesis *must* hold if the theory is true, then when the hypothesis is *not* supported we *must* doubt the truth of the theory.

Induction involves reasoning from the specific to the general, from a set of observations to a general conclusion. Here is an example:

Observation: Billy Joe, Leroy, Juan, and Betty Jane live in Dallas and each is a fan of the Dallas Cowboys.

Conclusion: People who live in Dallas are Cowboy fans.

Conclusions arrived at through induction are *empirical generalizations*. Even though they are derived from observations, however, empirical generalizations, like theories, can never be proven for all time. New observations may find exceptions or even strong evidence that the generalization is not true or no longer is true. In his famous treatise on induction and deduction, the British philosopher David Hume (1711–1776) noted that, even though we have observed a million crows, the empirical generalization "All crows are black" is forever vulnerable to the discovery of a white crow or to a mutation that results in yellow crows. Moreover, empirical generalizations based on only a few cases run a great risk of falsification or of at least being overstated. Given an adequate number of observations, the generalization about people who live in Dallas would need to be modified to assert only that *most* people who live in Dallas support the Cowboys—for some Dallas residents aren't interested in football and some support other teams.

SOURCES OF THEORIES

Where do theories come from? They come from our imaginations and our observations. Most theories are prompted by the desire to explain something we know exists or is taking place. Why are some societies less stratified than others? Why do some people join unusual religious movements? As social theorists reflect on such questions, they begin to sketch answers. Sometimes, they will move directly toward formulating these answers into a theory, but more often they will gather additional information before taking the step into theory. It undoubtedly would help in constructing a theory of stratification to know that there is very little stratification in hunting and gathering societies and a great deal of it in agrarian societies (Harris, 1979). It would undoubtedly help a sociologist of religion to know that most converts to unusual religious movements are young and relatively well educated (Stark and Bainbridge, 1985). However, when social scientists gather additional information as a preliminary to theory construction, they usually do not rely on casual observations. Usually, they rely on research.

Research results not only may require social scientists to abandon or modify theories, they also may prompt them to create new theories. This is especially true when research reveals surprises—things not previously recognized. Sometimes this occurs as a byproduct of research when observations made for one purpose turn up something no one was looking for or expected. Research during World War II on how satisfied American soldiers were with the speed of promotions accidentally discovered that satisfaction was far higher in units with the *slowest* rates of promotion (Stouffer, et al., 1949 I:256). This led the researchers to invent the principle of *relative deprivation*—that one feels more or less deprived on the basis of comparisons between one's own situation and that of comparable others. Thus, soldiers based their satisfaction with the speed of promotion on comparisons between their own progress and that of other soldiers in their unit. In units with rapid rates of promotion, the average soldier felt he was not keeping up *relative* to others. In units with few promotions, the average soldier was keeping up with nearly everyone else.

EXPLORATORY RESEARCH

Some discoveries are happy accidents, but social scientists often set out in hopes of discovering a surprise or serendipitous empirical generalization. In fact, some research doesn't involve hypothesis testing at all. Sometimes, social scientists engage in speculative and **exploratory research**, in which they make systematic observations of uncharted and little-known phenomena in order to get an initial sense of what is going on. For example, **Michael Jindra** (1994) made extensive observations of "Star Trek" fans, hoping to understand the basis for their intense participation in fan activities and organizations. One of his surprise findings was that, among trekkies, women greatly outnumber men and that this is especially true among the most active and prominent figures in the world of "Star Trek" fans. Because women also are known to be more active than men in religion, this led Jindra to explore the extent to which involvement in "Star Trek" serves as an alternative religion.

PURE, APPLIED, AND EVALUATION RESEARCH

As the example of exploratory research makes clear, not all social research is conducted to test theories. In fact, most is *not*. On the basis of its primary motivations, social research can be differentiated into three broad categories: *pure* (or basic) research, *applied* research, and *evaluation* research. However, these categories are not mutually exclusive, and it is possible for one research study to be all three. Moreover, each may involve theory testing.

Pure Research. The primary motive in **pure research** (or basic research) is directed by the desire to increase knowledge, without regard for potential practical applications. Such research usually involves theory testing and/or efforts to refine or extend theories.

Eva M. Hamberg and Thorleif Pettersson (1994) wanted to test a new theory about the effects of religious competition on religious participation. This theory predicts that rates of church membership and participation in religious activities will be higher where more churches compete for support. The traditional sociological view of competition was that it hurt religious participation. Hamberg and Pettersson based their study on the 284 municipalities of Sweden. Because Sweden has a very low rate of church attendance (about 6 percent attend in a given week) and because there are few denominations competing with the state-supported Church of Sweden (Lutheran), this posed an especially stringent test of the theory. Nevertheless, the findings very strongly supported the new theory: In municipalities where there was greater competition, there was more attendance.

While this study was regarded as a very important contribution to knowledge by sociologists of religion, it is difficult to see how the findings could be put to practical use. Churches aren't going to flock north to compete with the Church of Sweden simply because by doing so they would improve overall rates of church attendance in Sweden. This is, therefore, an example of pure research—done for the sake of knowledge alone.

Applied Research. The primary purpose of applied research is to serve practical needs. Applied research may or may not involve theory testing but most often does not.

Barbara Heyns (1978) was concerned about the way in which children from poor homes fell further behind in school each year. She suspected it might not be that they learned less than did children from more advantaged homes but that such children lost ground during the summer because they did nothing "educational" then. She formulated the hypothesis that the gap between the achievement scores of grade school kids from higher- and lower-income homes increases only during the summer. To test this hypothesis, she arranged to have achievement tests given to all Atlanta schoolchildren at the start of the fifth grade, the end of the fifth grade, and the start of the sixth grade. She found that family income had a negligible effect on increases in achievement scores during the fifth grade—between the fall and the spring. But the average scores of kids from lower-income homes actually declined over the summer, while the scores of kids from higher-income homes increased modestly between the spring and fall testing. Heyns traced these differences to the fact that kids from lower-income homes tended not to read during their summer vacations, while the kids from the higher-income families read a lot. She did not derive her hypothesis from a theory, and her findings do not contribute to our basic social scientific knowledge. However, her research was of immense practical importance and is, therefore, an excellent example of applied research. We now know, for example, that extending the school year might do much to mitigate the educational deficiencies associated with poverty.

Evaluation Research. Scientists do evaluation research to assess the effectiveness of a program, policy, product, or procedure, and it usually is commissioned by government agencies, businesses, or organizations such as schools, churches, or hospitals. If the program, policy, procedure, or product being evaluated is based on a theory, then evaluation studies also involve theory testing. But this usually is not the case.

Peter H. Rossi, Richard Berk, and **Kenneth J. Lenihan** (1980) were concerned about the very high numbers of persons who commit new crimes following their release from prison. They concluded that a solution for at least part of this problem might be to give people financial aid for a period following their release from prison to help them get reestablished. Officials in Georgia and Texas agreed to adopt this policy for a trial period but asked Rossi, Berk, and Lenihan to conduct research to evaluate its effectiveness. The trial began when the two states began to give weekly paychecks for six months to half of those released from prison (recipients were selected randomly) and nothing to the other half. At the end of their first year out of prison, 49 percent of those who did not receive checks had been rearrested. And of those who received checks, 49 percent also had been rearrested after a year. The program was a complete failure.

This evaluation research was extremely valuable in that it offered solid evidence that a very plausible program did not work and that the program's strategy should not be used to try to reduce the rate of reoffending. In doing so, it also qualified as applied research. But it offered us no insights into why people commit crimes or why doing time in prison does not deter people from new offenses. Thus, the study had an immediate practical payoff but added little or nothing to the stock of fundamental knowledge.

A HISTORICAL NOTE: WHEN SOCIAL SCIENTISTS WENT TO WAR

Given the reliance of social scientists on government funding and the reliance of government on social scientists for evaluation and applied research, it might be supposed that there has been a close link between social scientists and government agencies from the beginning. But, in fact, social research went on for decades before any appreciable government support was available. Social researchers struggled along as best they could. Then came Pearl Harbor and suddenly social research came of age.

Soon after the United States entered World War II, the War Department[2] was faced with the daunting task of rapidly training millions of young draftees. Someone suggested that social scientists conduct evaluation research on various training programs to help discover what worked best. So Samuel A. Stouffer, a Harvard sociologist, was recruited to organize the Research Branch of the Information and Education Division. Stouffer soon gathered a staff that included some of the most prominent social scientists in the country as well as a number of rising young stars—including several well-known women who joined the Research Branch as civilian employees.

---

[2] At that time, the Army and the Air Force were under the command of the War Department, while the Navy and Marines came under the Department of the Navy. Soon after World War II, these two departments were merged to become the Department of Defense.

With the vast resources of the War Department at their disposal and millions of soldiers to whom they could administer all sorts of questionnaires and psychological tests, social scientists had for the very first time the resources to do really large-scale studies utilizing proper sampling techniques. In addition, Stouffer brought in Carl Hovland to direct a section of the Bureau devoted to experimental research. The experimentalists conducted a number of famous experiments to evaluate motivational films such as Frank Capra's *Why We Fight* series (Capra also directed *It's a Wonderful Life*, the Christmas classic staring James Stewart).

By war's end, an incredible number and variety of major studies had been completed, and these soon appeared in four huge volumes (see Box). It would be impossible to exaggerate the substantive as well as methodological impact of this work on all of the social sciences—as noted earlier in the chapter, one of the discoveries led to the principle of relative deprivation. Not surprisingly, having served in the army with Sam Stouffer became a major status symbol, and "Sam's veterans" soon dominated the most famous sociology, political science, and psychology departments in the nation. Others became extremely influential figures in educational psychology and mass communications. The bonds formed among these scholars during their years of service also prompted them to form the American Association for Public Opinion Research (AAPOR) shortly after the war. Today, this organization brings together academic, governmental, and commercial survey researchers and publishes *Public Opinion Quarterly*.

PRIMARY PUBLICATIONS OF THE WAR DEPARTMENT'S RESEARCH BRANCH

✦ ✦ ✦ ✦ ✦

Stouffer, Samuel A., Edward A. Suchman, L. C. Devinney, Shirley A. Star, and Robin M. Williams, Jr. *The American Soldier: Adjustment During Army Life, vol. I.* Princeton, N.J.: Princeton University Press, 1949.

Stouffer, Samuel A., Arthur A. Lumsdaine, M. H. Lumsdaine, Robin M. Williams, Jr., M. Brewster Smith, Irving L. Janis, Shirley A. Star, and Leonard S. Cottrell, Jr. *The American Soldier: Combat and Its Aftermath, vol. II.* Princeton, N.J.: Princeton University Press, 1949.

Hovland, Carl, Arthur A. Lumsdaine, and Fred D. Sheffield. *Experiments in Mass Communications.* Princeton, N.J.: Princeton University Press, 1949.

Stouffer, Samuel A., Louis Guttman, Edward A. Suchman, Paul F. Lazarsfeld, Shirley A. Star, and John A. Clausen. *Measurement and Prediction.* Princeton, N.J.: Princeton University Press, 1950.

The most lasting impact of the Research Branch was that it forged a link between the government and the social sciences. Today, hundreds of millions of dollars are spent each year on social research by federal, state, and local government agencies. Some of the money goes to basic research. Considerably more goes to applied and to evaluation research. And it all began with the Research Branch.

CONCLUSION

Concepts and theories can have many pedigrees and a variety of inspirations, just as they can have many purposes and applications. But, following their creation, these mental constructs all sustain a single imperative: Theories exist to be tested, and the more *rigorous* the tests, the greater the probability that the theories are correct. Even if research does not involve theory testing, the same fundamental imperative must be observed. To qualify as social *science* rather than as opinions and hunches, theories must meet rigorous standards of observation and analysis. The remainder of this book explains these standards and how they can be fulfilled.

REVIEW GLOSSARY

- Scientific **concepts** are abstract terms that identify a class of "things" to be regarded as alike.

- An **abstraction** is an intellectual creation, a definition, existing only in our minds. Because scientific concepts are abstract they apply to all possible members of the class, all that have been, are, shall be, or could be.

- As used by modern philosophers of science, the **law of parsimony** (sometimes referred to as **Ockham's razor**) reads: *Theories always should attempt to explain the most with the least.* We should try to explain as much as possible with a theory that is as simple as possible. Applied to concepts, parsimony encourages greater abstraction.

- As Carl G. Hempel (1952) pointed out in his classic work on scientific concepts, it sometimes is claimed that "real" definitions somehow are an intrinsic part of the thing being defined, but this is folklore. Scientific definitions are not thought of as real in this sense at all, but are regarded as names that simply are *assigned* to something. To underscore this point, Hempel identified all scientific concepts as **nominal definitions**—that they are merely names.

- **Utility** is the ultimate test of all concepts, and utility means nothing more nor less than whether or not they turn out to be useful in constructing efficient theories. *What works better, is better!*

- Efficient concepts will have **clear boundaries**, delineations that eliminate ambiguity about what a concept does and does not include.

- A **tautology** is any statement that is true by definition as in President Calvin Coolidge's famous remark that "as more and more men are thrown out of work, unemployment results."

- **Theories** are abstract statements saying *why* and *how* some set of concepts are linked. Their purpose is to *explain* some portion of reality. Because theories contain concepts and concepts are abstractions, all scientific theories also are abstractions. That is, you can't see or touch a theory. There is one more vital feature of real theories: They must make empirical predictions and prohibitions—that is, it must be possible to say what sorts of observations would **falsify** the theory.

- **Empirical** means "observable through the senses."

- An **indicator** is an observable measure of a concept.

- A **hypothesis** tells us what to expect to observe when we examine relationships among indicators. If the hypothesis is derived from a theory, the theory specifies relationships among concepts and this is reflected in the hypothesis, which specifies the relationships to be observed among indicators. Even when hypotheses derive from hunches or common sense rather than from theories, they specify where we should look and what we ought to observe.

- **Correlation** means "to go together, to vary in unison." Correlations can be either positive or negative. If sales of ice cream rise when the temperature rises and fall when the temperature falls, there is a positive correlation. However, if the sales of down jackets fall when the temperature gets warmer and rise when it gets colder, there is a negative correlation.

- **Type one error** involves accepting that the predicted correlation exists, when in fact it does not.

- **Type two error** involves concluding that the predicted correlation does not exist, when in fact it does.

- **Null hypothesis** asserts that a hypothesis derived from a theory is false. That is, lacking persuasive evidence to the contrary, we accept the null hypothesis which states that there is no correlation among indicators of a theory.

- An **empirical generalization** is a summary statement based on empirical observations. "Men are more likely than women to favor capital punishment" is an empirical generalization summing up the results of many surveys.

- **Deduction** involves reasoning from the general to the specific—from a known (or assumed) principle to an unknown but observable conclusion. Put another way, when scientists use deductive logic they begin with a general, abstract premise or proposition and then show that less general, empirical predictions are implied by the statement according to the rules of logic.

- **Induction** involves reasoning from the specific to the general, from a set of observations to a general conclusion. Empirical generalizations are obtained through induction.

- **Exploratory research** occurs when social scientists make systematic observations of uncharted and little-known phenomena in order to get an initial sense of what is going on.

- **Pure research** (or basic research) is directed by the desire to increase knowledge, without regard for potential practical applications. Such research usually involves theory testing and/or efforts to refine or extend theories.

- The primary purpose of **applied research** is to serve practical needs. It may or may not involve theory testing but most often does not.

- **Evaluation research** is research conducted to assess the effectiveness of a program, policy, product, or procedure, and it usually is commissioned by government agencies, businesses, or organizations such as schools, churches, or hospitals. If the program, policy, procedure, or product being evaluated is based on a theory, then evaluation studies also involve theory testing. But this usually is not the case.

Steps in the Social Scientific Process

S ocial scientists observe the world from many angles and organize their activities in different ways. Nevertheless, it is possible to lay out a general step-by-
step description of how they proceed. Examination of these steps will give you a
very useful preview of the remainder of the book.

STEP 1: SELECTING A TOPIC

Research always starts with wondering. Social scientists tend to do a lot of
wondering about human behavior. For example, while watching a baseball game,
a social scientist might notice players exhibiting a variety of superstitions and
begin to wonder about the role of superstition in sports. Another social scientist
may be concerned about several friends who are single parents and are having a
lot of trouble with their kids. This might prompt reflection on whether children
suffer from being raised in a single-parent home.

As social scientists ask themselves such questions, the first step they are apt
to take is to find out what is already known that might apply. For example, they
may try to determine whether there are existing theories that suggest an answer.
Thus, the social scientist interested in superstitions and sports might discover the
classic theory about the uses of magic and superstition developed by an
anthropologist.

Early in the twentieth century, Bronislaw Malinowski spent many years
observing social life among the Trobriand Islanders in the South Pacific. These
people relied considerably on magical and superstitious practices, and this soon
attracted his attention. However, Malinowski was not content to merely describe
the magical practices of Trobriand Islanders. He wanted to explain why magic
existed and how it was used.

His first important insight came when he noticed that the people he was
observing did not resort to superstitious practices haphazardly. Sometimes they
used them, sometimes they did not. When there were weeds in their gardens or
when animals knocked down their fences, they "will have recourse not to magic,
but to work guided by knowledge and reason" (Malinowski, 1948:28). Nor did
they have recourse to magic when they paddled their boats on the calm, protected
lagoon. But when they took their boats beyond the lagoon out onto the open
ocean, or when their crops needed rain—when confronted by threats over which
they lacked objective means of control—"to control these influences and these
only [they] employ magic" (1948:29). Malinowski's theory asserts that

> *within any human group, magical and superstitious practices will be focused
> on those important activities over which people have the least control.*

STEP 2: FORMULATING A RESEARCH QUESTION

In addition to searching the theoretical literature, social scientists may be led by their questions to search the published research literature to see if others have worked on this question or on similar questions. For example, the social scientist interested in the effects of one-parent families would look to see what was already known. But, however social scientists proceed, their goal is to develop research questions that can be subjected to *empirical investigation*.

Whether derived from a theory, from the published research literature, from hunches or flashes of insight, a research question needs to be transformed into a statement, as a preliminary to the formulation of a hypothesis.

The social scientist interested in superstition in sports might proceed as follows. Baseball players have a great deal of control over their fielding and throwing. Errors are quite rare, and the average player will succeed in making fielding plays correctly at least 95 percent of the time. Hitting is something else. The best hitters fail two out of every three times at bat. Thus, it can be deduced from Malinowski's theory of magic that

> *baseball players will engage in many more superstitious practices about hitting than about fielding.*

The social scientist interested in the consequences of growing up in a one-parent family might proceed this way. The literature on single-parent families is complex and often contradictory. However, there is considerable agreement that children almost unavoidably are less supervised when there is only one parent around to keep track of them. Examination of the theoretical literature reveals that one of the most influential theories of crime and deviance—control theory—stresses the importance of supervision in limiting delinquent behavior (Gottfredson and Hirschi, 1990). Given the mutual implications of the research literature and control theory, it can be deduced that

> *children in one-parent families will be more likely to commit delinquent acts than will children in two-parent families.*

Notice that these statements are not yet researchable hypotheses because they still include some concepts rather than indicators. That is, we can observe children and single-parent families, but "delinquent acts" constitute an abstract concept. Similarly, we can see baseball players as well as hitting and fielding, but "superstitious practices" is a concept. Before we can state the research hypothesis, we must be sure all concepts are clearly defined and then we must select appropriate indicators of the concepts.

STEP 3: DEFINING THE CONCEPTS

Many different concepts are used by social researchers. Some concepts have a generally understood meaning and there is no need for further definition. In some cases, there are competing definitions and it is necessary to select among them. And, sometimes, the researcher develops a new concept and must be sure that others fully understand what it means. Here are definitions for two concepts that have a generally understood meaning:

> *Superstitious practices* are procedures used to influence events and for which there is no basis in evidence or logic to believe that the procedures work.

> *Delinquent acts* constitute behavior in violation of the law.

STEP 4: OPERATIONALIZING THE CONCEPTS

Once we know what we are talking about, it is time to seek effective and feasible indicators of our concepts. This process is called **operationalizing** the concepts.

Many of the concepts used by social scientists have standard indicators. For example, most studies of delinquency rely on self-reports—on asking adolescents questions about their past behavior. A huge amount of work has been devoted to developing valid and effective sets of items for use in this research, and most researchers rely on these items as their indicators of delinquency.

To operationalize some concepts requires originality. There are no standard indicators for superstitious practices. But it seems clear enough that indicators of superstition among baseball players would include such things as the use of charms or lucky objects (rabbit's foot, lucky medallion) and/or certain rituals and routines (Willie Mays always touched second base on his way in from center field) used by players to influence their success.

It is important to select indicators carefully because the way in which we operationalize our concepts can affect the results of our studies. And social scientists always should try to use the best possible indicators of their concepts. However, the best among *possible* indicators often are not the very best indicators—practical and moral considerations force compromises. For example, the very best way to study delinquency would be to place a sample of teenagers under 24-hour surveillance in order to find out what they actually do. But this would be far too expensive, even if there were no ethical objections to doing so.

STEP 5: FORMULATING THE HYPOTHESIS

Once researchers have indicators for all of their concepts, they can transform their research questions into testable hypotheses. Here are two examples:

Hypothesis: Baseball players will more often use lucky charms, rituals, and magical procedures when they attempt to hit than when they attempt to make a play in the field.

Hypothesis: Children growing up in one-parent homes will score higher on a delinquency questionnaire than will children growing up in two-parent households.

STEP 6: MAKING THE OBSERVATIONS

A hypothesis not only tells us what to expect to see, it also tells at whom or at what to look. Both of the hypotheses above direct our attention to the behavior of individuals, not families or teams.

It is at this stage in the research process that social scientists actually go out to look at the world—whether it is by direct observation or by use of information-gathering methods such as questionnaires, interviews, official records, and the like. For example, the social scientist interested in baseball might attend many games and carefully record all visible instances of reliance on superstitious practices and indicate whether each was associated with hitting or fielding. It also might be suitable to interview players about their superstitions and about when they use them.

Often, the appropriate data already exist so the researchers needn't seek more. Thus, the social scientist interested in one-parent families would not need to collect new data because there are many very large data sets already available, based on surveys of teenagers, that include measures of delinquency as well as details about the structures of their families. Therefore, anyone wishing to test that hypothesis could do so almost immediately.

STEP 7: ANALYZING THE DATA

Once the data have been collected or otherwise obtained, it is time to check out the hypothesis. For example, data on teenagers would be placed in a computer and the researcher would then compare teenagers living in one-parent homes with those in two-parent homes and see which group had higher delinquency scores. In similar fashion, the sociologist studying baseball would see whether there was more magic associated with hitting than with fielding. Making such comparisons is called *statistical analysis*. You will have many opportunities to analyze data during this course.

STEP 8: ASSESSING THE RESULTS

The first thing researchers want to know is whether their hypothesis (or set of hypotheses) was supported or rejected—whether the data turned out as predicted.

DATA ARE

◆　　◆　　◆　　◆　　◆

In science, observations are referred to as **data** whereas a single observation is a **datum**. That is, the word *data* is plural. People often slip and treat the word as singular as in "The data *is* ready." The correct sentence is "The data *are* ready." As a guide, you can mentally substitute the words *fact* and *facts* for *datum* and *data*—you wouldn't say, "These facts is unreliable."

If the hypothesis is supported, the researcher usually is able to wrap up the study rather rapidly by writing a report of the results for publication. However, when the hypothesis is not supported, and especially if the hypothesis was derived from a theory, social scientists face a much more extended process of assessment.

As part of such an assessment, researchers often will retrace their steps to see if each was done properly. Is the hypothesis really consistent with the theory, or did the deductive process slip? Are the concepts properly and clearly defined? Do the indicators really measure the concepts? Are these the correct units of analysis? Were the data properly collected? Was the analysis done correctly? If all of the answers are "yes," then the social scientist must suspect that something is wrong, or at least incomplete, in the theory itself.

At this point, some researchers publish their findings and let others attempt to discover what went wrong. Others attempt to reformulate the theory. But, if the findings are both important and surprising, as all failures of significant theories are, then it is certain that this is not the last word on the matter. Other people will do the research over again to be sure the results are trustworthy (see Step 10).

STEP 9: PUBLISHING THE FINDINGS

Most social scientists probably like to see their names in print, but that isn't the primary reason they publish articles and books based on their research. The fact is that an unpublished study might as well not have been done at all and, in an important sense, it wasn't. Research findings stuffed into a desk drawer or gathering dust on a shelf make no contribution to knowledge. All sciences are public, social activities—they exist primarily as a group of people who are engaged in common pursuits. And the primary foundation of such groups is the written word. Of course, scientists talk to one another and important findings often spread rapidly by word of mouth. But speech is imprecise and leaves no enduring record. It is through publications that members of a scientific community inform one another, and it is publications that link past members to current members and current members to future members. Moreover, publications must be sufficiently precise so that other scientists can *replicate* the research.

STEP 10: REPLICATION RESEARCH

Replication research is done by repeating previous research to check on the results. This is a very frequent and very important kind of research. Its importance stems from several factors that apply to all sciences, not just the social sciences.

The first factor is that a slightly different selection of indicators, or doing a study on a different set of units of analysis, at a different time or a different place might change the results. This is especially likely in social sciences because human beings are more variable than are molecules or bacteria.

A classic example of the importance of replication involved the first published test of the notion that religious commitment causes young people to be law abiding. Travis Hirschi and Rodney Stark (1969) found that among teenagers in Richmond, California, those who went to church and Sunday school were *not* less likely to commit delinquent acts than were teenagers who never went to church. Such unexpected findings attracted a great deal of interest, and even Hirschi and Stark wondered whether their results could be true. So several years later, the results of a new study conducted in Oregon were published (Burkett and White, 1974). The results were the same: Religious participation did not reduce delinquency.

But not even two studies were sufficient to convince social scientists and so replication research continued. Then, almost simultaneously, three new studies appeared, each reporting the missing religious effect. In Atlanta, students who went to church were much less likely to be delinquent than were nonattenders (Higgins and Albrecht, 1977). The same was true in Winnipeg (Linden and Currie, 1977) and in Utah (Albrecht, Chadwick, and Alcorn, 1977). Soon another study found religious kids were less delinquent in Arizona (Jensen and Erickson, 1979). Eventually, it was determined that religion does reduce delinquency except under special circumstances. These circumstances, which include unusually low levels of church membership in the community, are peculiar to the West Coast, where the first two studies—each of which failed to find a religious effect—were conducted (Stark, Kent, and Doyle, 1982; Stark, 1996).

The second factor giving importance to replication research is that random flukes can produce false findings. Replication will reveal such flukes because the chances of them happening again are slim.

The third factor is that research can be careless, incompetent, or dishonest. Unfortunately, no profession or occupation is without its fools and frauds; replication helps to weed out both.

Obviously, then, unpublished research or published reports that are very sketchy about how the research actually was done, are of little value.

CONCLUSION

This chapter offered an overview of the social research process. As such, it serves as an overview of the rest of the book. In succeeding chapters, we will retrace many of these steps in far greater detail. But, if the primary goal of the chapter was to offer a preview of coming attractions, a secondary goal was to set the tone for what is to come. A very important aim is to demonstrate not simply that research is useful and important, but also that it is fun and full of surprises.

REVIEW GLOSSARY

- The process of selecting indicators of concepts is called **operationalizing** the concepts.

- In science, observations are referred to as **data** whereas a single observation is a **datum**.

- **Replication research** is done by repeating previous research to check on the results.

3

Measurement

The first chapter clarified how indicators of concepts allow social scientists to test empirical predictions derived from theories. Thus, the actual observational part of research is based on indicators, for these are the empirical *measures* of the concepts. In this chapter, we examine basic principles involved in such measurement.

VARIABLES AND VARIATION

Social scientists try to explain what goes on in the world. More than that, they try to account for *variations* in what is going on. No one is much interested in the fact that all people have personalities, that all cities have crimes, or that all states have liquor consumption. What interests us is that people differ a lot in self-confidence, that some cities have far less crime than do others, and that some states have far higher rates of alcohol consumption and public drunkenness. The contrast here is between constants and variables.

CONSTANTS

Constants are characteristics or aspects of the things being studied that do not vary, but that take the same value.

All cities have crimes; therefore, having crimes is a constant, just as breathing is a characteristic of all living human beings. Constants often are of great use in science—the speed of light is a constant of great importance to physics. However, a constant cannot explain a variable. That is, something that does not change cannot cause other things to change.

VARIABLES

Variables are characteristics or aspects that take different values among the things being studied and thus can be said to vary.

Homicide rates vary across the 50 states, and therefore the homicide rate is a variable as are all differences among states, including the proportions of their populations who bowl, eat pizza, or drive pickup trucks. At the individual level, all ways in which people differ are variables: height, weight, religion, hobbies, voting preferences, and the like.

The primary task of social science is to *explain variation*. Social scientists do this by trying to *discover connections among variables*. And that is, of course, precisely what hypotheses do: They predict such connections.

In the second chapter, all of the indicators we examined were variables. Family structure varies: Some families have only one parent, and some have two. Baseball players differ in how much they rely on superstition, and the amount of superstition of players varies from situation to situation.

Since social scientific hypotheses nearly always consist of variables, when social scientists talk about research they nearly always refer to their variables, not to their indicators, as in:

"Age is by far the most powerful variable influencing . . . ."

"Maybe you haven't measured that variable correctly."

"Why don't you try some other variables?"

When we call something a variable, we are claiming that it takes at least *two* values. Gender is thus a variable, as it takes the values of male or female. Education, measured as years of schooling, is another variable taking values from zero to in excess of 20 years. Not only do variables differ a lot in their amount of variation, but the same variable may take many values in some settings and few or none in other settings. For example, in the general population there is a great deal of variation among people on the basis of age. Among high school students, there is little variation in age.

To explore variation further, examine this question asked in the 1993 General Social Survey of American adults:

During the last 12 months did you go to an auto, stock car, or motorcycle race?

Yes   16%
No   84%

Answers to this question constitute a variable, since answers can take two values. However, the variation is quite limited because so many people chose the same answer. There would be greater variation had respondents split evenly, with 50 percent attending and 50 percent not. On the other hand, there would have been no variation at all if *everyone* had attended or if no one had. Simply because the question offers people two choices does not qualify it as a variable. If everyone chooses the same answer, then the data represent a constant—something true of everyone. Consequently, there is considerably more variation in responses to this question asked in the same survey:

In the last 12 months did you participate in any sports activity such as softball, basketball, swimming, golf, bowling, skiing, or tennis?

Yes   56%
No   44%

However, both questions minimize variation because they offer only two options. Variation increases greatly if there are more categories (as long as people are well distributed across these categories). This example also comes from the 1993 General Social Survey:

How do you like jazz music?

Like it very much	16%
Like it	35%
Mixed	25%
Dislike it	19%
Dislike it very much	5%

Variables based on survey data often have only limited variation in comparison with variables based on such data as state crime rates or the percentage of each state's population who own guns.

Variables: Levels of Measurement

Social scientists use many kinds of variables. An important aspect of variables is their measurement properties. Four levels of measurement are identified.

Nominal or Categorical

Nominal or **categorical** variables sort cases into categories. Gender is a nominal variable; cases can be separated into one of two categories: male or female. However, the categories of a nominal variable lack intrinsic *order*. Thus, although people can be separated into the categories of male and female, it is impossible to say that one person has more gender than another—the categories lack any inherent order. Race, religion, region, and marital status are other commonly used categorical variables. They often take many categories, but there is no way to order them in terms of more or less, high or low.

Ordinal

Ordinal variables, not surprisingly, have the property of order. Not only can cases be sorted into categories, but the categories have an inherent order.

The survey question about jazz is an ordinal variable. People can be ranked along a continuum from liking jazz very much to disliking it very much. The same is true of the items on sports participation and on going to races, *even though they take only two values*. This is because these categories reflect quantity. People who have gone to a race in the past year have done so more often than those who have not. Those who have participated in sports at least once in the past year have done so more often than those who have not. The items could undoubtedly be improved by increasing the number of answer categories to distinguish among those who engage in these behaviors frequently and those who do so less frequently. But both versions of the variable qualify as ordinal.

Interval

Interval variables are a more precise form of ordinal variable in that the gaps or intervals between categories are of equal quantity.

We don't know if the difference between liking jazz very much and just liking it is the same size as the difference between disliking it very much and just disliking it. In contrast, when the intervals between categories are known to be of equal size, then we are dealing with an interval variable. Temperature measured on the Fahrenheit scale is an interval variable. We know that the difference in temperature between 20 degrees and 30 degrees is the same as the difference between 80 degrees and 90 degrees. True interval variables are extremely rare, since virtually all such measures also qualify as ratio variables.

RATIO

Ratio variables not only have equal intervals between categories, but also have meaningful zero points. A zero point is a point at which there is a total absence of the property being measured. Weight, for example, is a ratio measure. We know not only that the distance between categories (pounds, for example) is equal, but also that zero means the absence of weight. Temperature, when measured on the absolute scale, is also a ratio scale as absolute zero indicates the absolute lack of heat. Ratio scales allow us to express comparisons between cases as ratios—to say that a person weighing 220 pounds is twice as heavy as a person weighing 110 or that someone age 40 is twice as old as someone age 20.

As used by social scientists, however, variables based on individuals may not be true interval or ratio measures. In terms of how people respond to one another, among adults a difference of a foot of height between a person 4 feet tall and one 5 feet is probably perceived as much greater than the difference between a person 5 feet and a person 6 feet tall. That is, the first comparison is between a person in the normal height range and a person who is not, whereas both persons in the second comparison are within the normal range. The maturational and social significance of years of age also differs at various points in the life cycle. In terms of physical maturation, the interval between 2 and 3 is undoubtedly far larger than the interval between 25 and 26. Socially, the difference between 20 and 21 is probably greater than that between 30 and 31 or between 10 and 11. And, while we say that a person earning $100,000 a year has an income four times that of a person earning $25,000, it is clear that the additional dollars of income between $75,000 and $100,000 are not nearly so valuable to the individual earner as are those between $0 and $25,000 and, therefore, that the "ratio" property of this measure is not real.

In contrast, social variables based not on individuals, but on larger populations, probably often *do fully qualify* as interval and ratio measures. It probably is legitimate to say that the child mortality rate of Russia is twice as high as the rate of Italy or that the burglary rate in the Netherlands is twice that of Canada.

UNITS OF ANALYSIS

To test the hypothesis that social class is related to political conservatism, we would need to select an appropriate sample of people and then determine their incomes and whether they subscribe to *The National Review* (a magazine of conservative opinion). But suppose we changed the hypothesis slightly to predict that in *places* where income levels are higher, subscription rates for *The National Review* also will be higher. We can't test this hypothesis merely by collecting data on individuals. Instead, we need data based on cities, states, or some other varieties of "places."

Social scientists base their research on a variety of **units of analysis**—the "things" a hypothesis directs us to observe.

Particular units within the set defined as the units of analysis also are referred to as *cases*. Often, these "things" are individual human beings, and thus people are the units of analysis and each one is a case. Observations based on individual units of analysis often are referred to as *individual* data. But social scientists often base their observations on larger units. Thus, our second hypothesis directed us to observe large aggregates of people constituting places. Thus, places are the units of analysis and each place is a case. Units of analysis based on larger units often are referred to as *aggregate* data. Some examples will be helpful:

Individuals as the units of analysis:

Thomas C. Wilson (1991) wished to test the hypothesis that by moving from one community to another, *people* gain experience with cultural diversity and therefore become more tolerant than those who remain in the place where they were born. To find out, he examined data based on interviews with a national sample consisting of 9,276 American adults and compared the "movers" with the "stayers," finding that the former are indeed more tolerant of people holding controversial views.

Married Couples as the units of analysis:

Lynn K. White and **John N. Edwards** (1990) interviewed a national sample of married *couples* repeatedly over a number of years. They found that couples tended to express considerably greater happiness and satisfaction with their marriages *after* their last child had moved out. They called this the "postlaunch honeymoon."

Basketball Teams as the units of analysis:

Paul T. Schollaert and **Donald Hugh Smith** (1987) collected data on each of the *teams* in the National Basketball League to test the hypothesis that fan support is influenced by the racial composition of the team. If it is true that fans (most of whom are white) will not support teams that have too many African-American players, as many have claimed, then teams with more white players ought to have better attendance. The results showed that only one factor influenced attendance, and it wasn't race. It was winning and losing.

States and Provinces as the units of analysis:

Rodney Stark and **William Sims Bainbridge** (1985) wanted to determine the relative strength of occult and psychic activities in various parts of the United States and Canada. To do so, they obtained circulation figures for various magazines devoted to occult and psychic phenomena. Next, they examined the *Yellow Pages* of the phone books for the United States and Canada and counted the number of listings for astrologers. Finally, they obtained a North American directory of New Age stores and restaurants. These data were transformed into rates for each of the 50 *states* and for the 10 *provinces* of Canada. Data from all these sources gave the same geographic results: In both Canada and the United States, occult and psychic activities are highly concentrated in the Far West.

Nations as the units of analysis:

Katherine Trent and **Scott J. South** (1989) suspected that divorce rates rise when a higher proportion of women work outside the home. They collected data on 66 *nations* and found that their suspicions were supported—the higher the percentage of women in a nation's labor force, the higher the nation's divorce rate. They also discovered that divorce is higher to the extent that women outnumber men—that where men are in relatively short supply people are less likely to remain married.

Court Cases as the units of analysis:

Paul Burstein (1991) collected data on all appellate court cases decided between 1963 and 1985 involving suits based on the federal equal employment opportunity laws—a total of 2,081 cases. He found that suits were decided in favor of the plaintiffs far more often when the federal government joined the lawsuit on the side of the plaintiffs by filing an amicus brief.

Stage Plays as the units of analysis:

Mabel Berezin (1994) wished to determine the extent to which the Italian theater responded during the 1930s to Benito Mussolini's efforts to produce a new, fascist national culture. She examined 354 new plays that appeared on the Italian stage between 1934 and 1940. She found that only 5 percent "had themes that could be characterized as explicitly fascist."

AGGREGATE DATA AND THE ECOLOGICAL FALLACY

Since the unit of analysis is the thing being studied, observations gathered during research describe that unit, not components of the unit. That is, when we conduct interviews with a national sample of individuals, we can then accurately describe individuals. We can say that people with higher incomes are more likely to read *The National Review*. When we gather data on states, however, we cannot accurately describe individuals living within a state, because to create such data we have aggregated the behavior of individuals to create collective variables.

To aggregate is to put together; hence **aggregate data** refer to data based on units of analysis larger than the individual. Because, in a sense, aggregate data describe the social reality surrounding the individual, such data also often are referred to as **ecological data**.

Suppose we collect data from residents of each state, including their annual income and whether or not they subscribe to *The National Review*. Next, we calculate the average income of residents and the circulation rate of *The National Review*—the number of subscribers per 1,000 citizens, for example. What we have done is to aggregate individual data to produce variables describing states, not individuals. Individuals have incomes, not average incomes, and individuals either read or do not read a magazine, they don't have circulation rates.

It is very important not to forget what units we actually are describing when we interpret research findings. If we discover that circulation rates for *The National Review* are positively correlated with rates of average income, we can safely say that this conservative magazine sells better in wealthier *states*. We cannot say with certainty that the higher their incomes, the more apt *people* are to subscribe. We can strongly suspect that this is the case, but we cannot assume it to be true unless we find such a correlation when individuals are the units of analysis. The reason we cannot automatically apply aggregate results to individuals is that it *could* be that it is poor people who read this magazine, but that they are more likely to do so in places having higher average incomes. That sounds strange because it is unlikely—but it is possible.

When we assume that findings based on aggregate or ecological data apply to individuals, we are committing the **ecological fallacy**.

Let's consider a well-known example. Using states as his units of analysis, William Robinson (1950) found that during the 1930s there was a positive relationship between the literacy rate and the percentage of the population that was foreign born. While this was undoubtedly true of states, to conclude from this that foreign-born residents of the United States had a higher literacy rate than native-born Americans was false—the literacy rate actually was lower among the foreign born. What was going on? The American South was extremely poor in those days, and literacy rates were extremely low in southern states. Few immigrants settled in the South—there being few industries and no vacant farm land to draw them. Consequently, most immigrants settled in the wealthier states of the North and West where literacy rates among the native born were high, resulting in high literacy rates for these states despite their influx of immigrants.

Too much can be made of the ecological fallacy. While we must take care not to commit it, the fact remains that correlations at the aggregate level usually *do* reflect what would be found at the individual level.

THE QUALITY OF MEASURES

Whenever we use a variable, we must be concerned about whether it is a *reliable* measure of a concept and whether it *validly* measures that concept.

RELIABILITY

A variable is **reliable** if it is consistent—if repeated observations give similar results.

During their years in school, students take many standardized tests designed to measure their abilities and achievements. What if a group of students took a reading ability test three weeks in a row and their scores were extremely different each time? Obviously, their actual reading skills could hardly change at all during a three-week period, so the test results would be meaningless. Such a test would be discarded as unreliable, for failing to yield consistent measurements.

The first law of carpentry is to measure everything twice. While a tape measure does not stretch and is perfectly reliable, the person using the tape measure may make a mistake. Thus, if a carpenter obtains conflicting measurements, he or she will measure a third time. In science, we use this basic principle of multiple measurements to assess the reliability of the measure. If we measure a set of cases twice using the same technique, we can use the correlation between the two sets of measures to assess reliability of the measurement technique. The higher the correlation between the two measures, the higher the reliability of the measures. While there are several techniques for assessing reliability, each is based on this underlying idea.

A great deal of comparative research is based on data sets in which premodern societies are the units of analysis. One of these is known as the Standard Cross-Cultural Sample (see Chapter 6) and consists of 186 societies. Many of the cultures described in these materials no longer exist, so to create new variables it is necessary to work from written materials which are often the field reports written by anthropologists. Thus, when **Gwen J. Broude** and **Sarah J. Greene** (1976) of Harvard University wanted to study variations in sexual attitudes and practices across these societies, they had to transform written accounts into ordinal variables having numerical values. For example, they wished to create a variable measuring the frequency of premarital sex among females. Thus, each society was assigned to one of the these four categories:

1 = Universal or almost universal: almost all females engage in premarital sex.
2 = Moderate: not uncommon for females to engage in premarital sex.
3 = Occasional: some females engage in premarital sex, but this is not common or typical.
4 = Uncommon: females rarely or never engage in premarital sex.

Obviously, such coding is a difficult task; even a careful coder may misread the materials or fail to be equally careful with each case, thus contributing to the unreliability of the measure. To prevent this, Broude and Greene (1976:411) did as follows:

> *All societies were rated independently by the two authors on all codes. All ratings were checked for agreement between the two judges. Disagreements were discussed, each coder citing evidence in the [original source documents] that led to her rating. The final rating for each code on each society was a product either of initial agreement or of consensus between the two coders after discussion; if no consensus could be reached, the rating was omitted.*

Inter-Rater Reliability.   The approach, used by Broude and Greene, is referred to as **inter-rater reliability** and is virtually the rule for coding qualitative data.

More than one rater or coder is used for each case, and the correlation between, or among, coders provides an estimate of the reliability of the resulting measure. Thus, for example, photographs of children were ranked according to attractiveness by a number of independent raters, and the reliability of the ratings was established by a very high level of agreement among the judges (Dion, 1972). Or, as will be discussed in greater detail later in this chapter, 16 experts on American religion were used to rank a number of American religious denominations in terms of strictness (Iannaccone, 1994).

Test-Retest Reliability.   Another common method of assessing the reliability of a measure is **test-retest reliability**. The same cases are measured at two different times, and the correlation between the two scores is the estimate of the reliability of the measure. One problem with this method is that differences across time may reflect a true change rather than a lack of reliability. For example, in testing reading ability, the longer the period between the two measures, the more likely that differences reflect changes in ability rather than random errors. So this correlation may underestimate the degree of reliability. On the other hand, if the period is too short, individuals may remember their answers from the earlier test and simply give the same answer. In this situation, the correlation would overestimate the degree of reliability.

Alternate Forms.   This second problem can be solved by using **alternate forms**. Suppose that the test of reading ability had 200 questions and each question was assumed to have the same level of difficulty. You could divide the questions so that you had two tests, each with 100 questions. You could administer each test to the same individuals at different times, and the correlation between the scores would be the reliability of the variable. You could then use either of the tests to measure reading.

Split Halves.   The **split halves** method is the same as alternate forms reliability except that both forms are administered in one test. The correlation between the scores on the two halves is used as a measure of reliability. This technique has the advantage that the test needs to be administered only once; assessing reliability using either test-retest or alternate forms requires testing the cases at two different times.

Internal Consistency.   The split halves technique has the problem that the correlation between the two halves will depend to some extent upon which items are placed in which half. Suppose instead of actually splitting the items into two groups, we could find the correlation for each possible way of dividing the items into two equal groups. We could then use the average correlation as the measure of reliability. This is called the **internal consistency** method. **Cronbach's alpha** (Cronbach, 1951) is by far the most popular method based on this approach; the value of alpha is affected by both the number of items and the correlations among the items. Thus, if the average inter-item correlation remains the same, adding more items will increase the reliability of the variable. Similarly, a variable will be more reliable if the items on which it is based are limited to those having higher average intercorrelations.

Each of these methods of measuring reliability will result in a number between 0 and 1. One indicates "perfect" reliability; generally, a value of 0.7 or higher indicates that the measure is sufficiently reliable for use. Measures of reliability are designed to estimate only the effect of random errors, effects caused by misreading questions, mismarking answers, and so on. Systematic errors are not detected by these methods. For example, if everyone subtracts 5 years when reporting age, these errors will not be reflected in the reliability estimate.

In fact, in social science research, variables are seldom tested for reliability. There are several reasons for this. First, for many variables, reliability is not an important problem; self-reported sex, race, religion, education, and so on can safely be assumed to be sufficiently reliable for use in research. Second, many measures in social science are crude ordinal measures; reliability may be the least of the problems in using such measures. For example, reporting a reliability of 0.72 implies a high level of precision in measurement; this seems inappropriate when cases are simply sorted into three or four categories. Finally, developing measures and assessing the reliability of such measures is an expensive and time-consuming process. Creators of standardized tests can spend millions of dollars developing a new test. As consensus about the meaning and importance of certain concepts develops, measures of those concepts will improve.

VALIDITY

A variable can be reliable and yield consistent measurements but still lack validity. That is, a reliable measure still may be measuring the wrong thing. A variable is **valid** if it actually measures the concept it is meant to measure.

Does agreement with the statement "Most people will cheat if they think they can get away with it?" really measure moral cynicism? There are four common methods of assessing the validity of a measure.

Face Validity. **Face validity** is the most common basis for establishing validity. It relies on common sense—to conclude that, *on the face of it*, the variable obviously measures the concept. We are content to accept that income is a valid measure of social class. We are equally willing to accept it as self-evident that a question such as "Do you agree or disagree that there ought to be a law against marriages between persons of different races?" is a valid measure of prejudice. However, the validity of some variables is less obvious.

Convergent Validity. A somewhat more stringent test of validity is based on the principle that valid measures of the same concept must be correlated. This is called **convergent validity** because the indicators converge on a single, underlying, empirical base that represents the concept. Thus, if you have a set of variables, each having face validity and each being an independent measure, you may test them for validity by examining the correlation matrix. It can be argued that if most of the variables display high intercorrelations, any variables that do not can be discarded as invalid.

Criterion Validity.   The most stringent test of validity—**criterion validity**—is based on comparing a particular indicator with another that is *known to be valid* and which can, therefore, serve as a **criterion**. There are stronger and weaker forms of criteria. In survey research, for example, a variable having considerable face validity will serve as a criterion to validate other variables thought to measure the same concept. The case for validity rests upon demonstrating that each of the other variables is highly *correlated* with the criterion variable and that in analysis one can be *substituted* for the other without changing the results.

The stronger form of criterion validity involves a criterion of obvious validity that has been measured independently from the variables to be validated. Here delinquency research offers an excellent example. With enough money and effort, it is possible to check the local police and juvenile court records to see if what teenagers report about their behavior matches up with their records (or lack thereof). Or you can interview each teenager's friends and see if their reports of the teenager's behavior matches with his or her self-reports. In other instances, juveniles have been reinterviewed about their behavior while hooked up to lie detectors and the results compared with their initial responses. Thus, official records, friends' reports, and lie detector results can serve as criteria for validating self-report delinquency questionnaires. Each of these criteria has been used several times to successfully validate self-report delinquency measures (Hindelang, Hirschi, and Weis, 1981). Therefore, researchers are able to trust the inexpensive self-report technique to study delinquency.

Construct Validity.   Some concepts include a number of assumptions concerning the phenomena they are meant to isolate and identify. In such cases, it often is possible to test the validity of a measure by seeing if it meets these under-lying assumptions. This test is known as **construct validity**, defined by Carmines and Zeller (1979:23) as "the extent to which a particular measure relates to other measures consistent with theoretically derived hypotheses concerning the concepts (or constructs) that are being measured."

Consider IQ tests. In 1906, when the French psychologist Alfred Binet was assigned the task of distinguishing between Parisian schoolchildren who lacked the ability to learn and those who were not learning because of lack of effort, he began with a concept he called *intelligence*. In defining this concept, Binet built certain assumptions about the phenomenon of intelligence into the concept itself. First, he postulated that people have an innate, inborn capacity for learning. From this it followed that the greater a person's capacity to learn, the more the person should have learned over a given period of time. Second, he proposed that the distribution of intelligence within a general population approximates the normal curve. To create his famous test, Binet began by using a great many questions in his preliminary testing, selecting only those that clearly differentiated children on the basis of age and which were normally distributed. The finished test could claim validity on the grounds that it was correlated with age and that scores were normally distributed as would be required of any measure of intelligence as Binet has conceptualized this phenomenon.

Cross-Case Comparability.   Valid measures must be **comparable across cases**. As we will examine at length in Chapter 7, survey researchers must write questions that will be understood the same way by all (or nearly all) respondents. Otherwise, some respondents will be answering one question while others will be answering a different question. As a result, their responses cannot be compared meaningfully. The most severe problems of this sort occur in cross-national research in which the same questions are asked in several different languages.

Consider the problem researchers face at Statistics Canada when they conduct national survey studies of Canadians. Since many of the interviews will be conducted in English and many others will be conducted in French, researchers must make sure that the meaning of the questions hasn't been changed in the translation. Several methods are used by researchers faced with the need to equate questions in different languages. One technique is to write the questions in one language, have someone translate the questions into the second language, and then have this translation evaluated by others having command of both languages. Another technique is to have one person translate the items into the second language and then have someone else translate them back into the first language and then see how similar the original question is to the one translated back.

When such safeguards are not taken the results may be defective, if often hilarious. Consider efforts by Americans to use a questionnaire to gather data about student riots and demonstrations in Japan. The original questionnaire had been developed in the late 1960s when such goings-on were fairly common on U.S. campuses. Having translated the questionnaire into Japanese, the researchers sent a copy to every dean of every Japanese university. The questionnaire included many questions about student demonstrations and the administration's response including this question: "How many students were suspended for their role in demonstrations?" Every Japanese dean replied, "None." But the researchers knew that, in fact, many students had been suspended from Japanese universities. So they consulted another translator. She read the question as it was stated in Japanese and began to laugh uncontrollably. In Japanese, the question asked, "How many students were hanged for their role in demonstrations?"

When nations are used as the units of analysis, seemingly comparable statistics may be based on different data in different countries. For example, some nations, such as France, include attempted homicides in their published homicide rate, whereas others, including the United States, do not, making comparison of the French and American rates very misleading.

Even within a particular society, the meaning of the measure may differ across cases. Different respondents may interpret questions differently based on their own cultural and personal background. For example, a Southerner may understand the word *grits* to mean a "coarse hominy, typically served with breakfast," but someone from New England may understand it to mean that a person grinds his or her teeth.

A Variety of Measures

The richness and diversity of social science is perhaps nowhere more evident than in the variety of measures that researchers use. Even a concept as simple as gender can be measured in different ways: A respondent can report his or her sex, an observer could code the sex of the individual, another researcher could code sex from the individual's birth certificate, and so on.

Chapters 6 through 11 will offer details about measures and various measurement techniques within particular research designs. But it will be useful for you to get an overview of the variety of ways in which concepts can be measured.

Survey Questions and Indexes

Surveys are the most common method for obtaining information in social science, and most of the research reports in the popular media are based on survey data. In addition to asking about characteristics of respondents such as sex, race, and religion, surveys usually include questions about attitudes and opinion and self-reports of behavior such as drinking, going to stock car races, attending church, reading, voting, and so forth. Sometimes it seems that almost every conceivable human attitude, behavior, and opinion has been included in one survey or another. Chapter 7 will explore the best ways to phrase survey questions and some of the more effective formats.

Here let's take a quick look at using questions as a measurement tool. The following question asks about allowing an atheist to give a speech:

> There are always some people whose ideas are considered bad or dangerous by other people. For instance, somebody who is against all churches and religion. If such a person wanted to make a speech in your (city/town/community) against churches and religion, should he be allowed to speak or not?
>
> 1 Yes, allowed to speak
> 0 Not allowed

It would be reasonable to conclude that this item is an indicator of the individual's attitude toward freedom of speech (a concept) and to use it to sort respondents into those who support freedom of speech and those who don't.

However, suppose we had a number of other similar questions all of which were designed as indicators of attitudes toward freedom of speech. We might be able to improve the accuracy of a measure of free speech by *combining* them into a single measure. It is entirely appropriate to do this so long as we are sure that each variable measures the *same* concept. Obviously, there is no point in combining multiple measures of gender, race, or age. But, when social scientists are concerned about accurately measuring prejudice or support for free speech, it often is preferable to combine a number of similar measures than to rely on only one. Combinations of variables based on survey data often are referred to as **indexes** (or **indices**).

When social scientists create indexes based on individual-level data (such as survey questions), they typically proceed simply by adding the values of some set of indicators. Such an index is based on the assumption that *more* measures of the same thing will yield more sensitive measurements. Suppose we wanted to combine the earlier item with each of the following two items, often included in the U.S. General Social Survey, to create an index of support for free speech:

> Or consider a person who believes blacks are genetically inferior. If such a person wanted to make a speech in your (city/town/community) claiming blacks are inferior, should he be allowed to speak or not?
>
>   1 Yes, allowed to speak
>   0 Not allowed
>
> Or consider a person who advocated doing away with elections and letting the military run the country. If such a person wanted to make a speech in your (city/town/community), should he be allowed to speak or not?
>
>   1 Yes, allowed to speak
>   0 Not allowed

The logic of combining variables to create an index is precisely the same as is used in the tests teachers use to grade their students. Teachers obtain a grade by adding the correct answers on each student's test. To construct an index, social scientists also count the "correct" answers—those representing the presence of the quality being assessed. In this case a social scientist would assign the value 1 to each "Yes, allowed to speak" response (those in favor of free speech) and assign the value 0 to each "Not allowed" response. When the values were added, each respondent would have a score. These scores represent an ordinal measure of support for free speech ranging from 0 through 3.

Indexes are not limited to only a few items. Some attitude indexes include dozens of items. But the point of creating them is always the same: accurate measurement. We will undoubtedly more accurately classify people in terms of their support for free speech if we measure their views with three items instead of one item as long as these three items *do* measure the same thing.

In addition, by creating an index we often are able to *increase variation*. Each of the above items is a variable taking only two values; the index increases that to four. However, an index might not actually increase variation. Suppose people either supported free speech in all three instances or rejected it in all three. That would produce an index having only the values 0 and 3. As it turns out, people do not score only 0 or 3 on this index. They quite evenly spread out across all four values.

A potential problem in creating indexes is multidimensionality. The item on page 44 might have measured only the individual's attitude toward atheists and have had little or nothing to do with support of freedom of speech. When we combined this item with the two items that follow it, we assumed that each item was a measure of support for free speech. If this is not the case, we may encounter the problem of multidimensionality of trying to measure two or more dimensions, or concepts, at the same time. If an index taps multiple dimensions, there is no way to interpret the results; are we measuring attitudes toward freedom of speech, toward atheists, or both? Even a single question can tap two or more different dimensions if respondents have different interpretations of the question: Such an item lacks cross-case comparability.

SCALES

Index construction usually produces a variable which appears to be valid and reliable; however, there is no underlying measurement model which justifies the construction of indexes (other than tests for reliability). For example, when we construct an index, we implicitly assume that each question contributes the same amount to the final score, but we have neither a justification for this assumption nor the means to test whether it is valid. In an attempt to produce better measures, social scientists have developed various scaling techniques.

Scales are measures created by combining variables using an explicit measurement model. Such models typically include the following three elements:

1. assumptions about the nature of the component questions, or variables
2. rules for combining the set of questions, or variables, to produce the measure
3. techniques for assessing the quality of the resulting measure, or scale

Various scaling techniques will differ in these elements, and there are many such techniques. Here we will examine Guttman scaling, one of the simplest techniques, to see how such scales differ from indexes.

When researchers score an index, they are unconcerned about how people earn a particular score—about their *pattern* of responses. Thus, in the example above, they treat all scores of 2 as equal regardless of which two items drew answers supportive of free speech. But, when social scientists construct scales according to the method developed by Louis Guttman they do pay attention to patterns of response and attempt to combine items according to a set of assumptions about these patterns. Guttman scaling, or scalogram analysis, is seldom used today,[1] but its logic is very easy to understand. This technique assumes the items, or indicators, function in a "cumulative" manner. For example, the following set of items is cumulative:

    a. I am over 5 feet tall.
    b. I am over 5 feet 4 inches tall.
    c. I am over 5 feet 8 inches tall.
    d. I am over 6 feet tall.
    e. I am over 6 feet 4 inches tall.

---

[1] At one time all leading statistical analysis programs included a Guttman scaling routine; today none does.

If you agree with a particular item, you *must have* agreed with all items above it. If the items in a scale are truly cumulative, and if subjects answer correctly, only six "scale types" are possible for a five-item scale: those who agree with all five items (scored 5), those who agree with the first four items (scored 4), those who agree with the first three (scored 3), and so on down to those who agreed with none (scored 0). Individuals are scored at the highest item with which they agree; if the scale functions perfectly, then this is the same as counting the number of items endorsed.

The social distance scale, originated by Emory S. Bogardus (1924), is an example of a Guttman scale. *Social distance* conceptualizes an aspect of prejudice. It refers to the closeness or intimacy of relationships a member of one racial or ethnic group is willing to extend to members of another racial or ethnic group. Suppose we were conducting a national survey in Belfast, the capital of Northern Ireland, and wished to measure the intensity of anti-Catholic and anti-Protestant sentiments. We could do so by using of a set of questions designed to measure each respondent's desire to maintain social distance from members of the other group. We could start with the most extreme amount of distance, and each next question could move the other group closer. The items asked of Protestants would be like these (Catholics would be asked similar items concerning Protestants):

> Thinking now about Catholics, do you think they ought to be allowed to live in this country or should they all have to move south to the Republic?
>
> What about here in Belfast? Do you think Catholics ought to stick to their own neighborhoods or do you think they should be free to live anywhere they wish?
>
> How would you feel about having Catholics living on your block?
>
> How about Catholics becoming regulars at your pub?
>
> Would you be willing to invite Catholics to have dinner in your home?
>
> How would you feel if a member of your family married a Catholic?

Presumably, persons who want to ship the Catholics out of the country would reject all items permitting closer contact, and Guttman scaling also would assume that those who would welcome a Catholic into the family would accept all more distant modes of contact.

In Guttman scaling, the order of the items is determined by the responses. If the assumptions are correct, the item with the largest number of positive responses should be the "lowest" item, the item with the second largest number of positive responses should be next, and so on. The item with the smallest number of positive responses should be the "highest" item. This ranking determines what "scale types" should be observed in the data. Unfortunately, in most applications of this technique, nonscale types—individuals who have patterns of responses other than those anticipated—will appear. That is, people aren't nearly as predictable or consistent as social scientists might like them to be. In our Belfast survey, it would not be unusual to find many people who would not object to having Catholics to dinner but still would not want them living on their block.

Others might not mind having *a* Catholic in the family but would not want a *group* of them frequenting their favorite pub. If there are a large number of nonscale types, then one or more assumptions of the measurement model has been violated and the scale is not acceptable. The *index of reproducibility* provides a measure of the quality of the scale.

Let's review how a Guttman scale differs from a simple index. First, in a Guttman scale, we have made an assumption about the nature of the component items—responses to the items will be cumulative. Second, we have a method for constructing a scale—individuals are placed at the highest item with which they agree. Third, we have a means of evaluating how well the final scale fits this model.

There are many other scaling techniques, each of which uses different assumptions and different procedures. However, researchers in most of the social sciences make far less use of scales today than they did 20 or 30 years ago. Though the decline in scaling probably has many causes, an important one is that the development of a scale is time consuming and expensive and the resulting scales have only a limited payoff for most social scientific uses. Since most scales are designed to measure attitudes or other psychological states, the use of scales probably fluctuates with the popularity of studies of attitudes; this area was very popular in the 1950s and 1960s but is far less so today except in psychology and some areas of educational research. Another reason scales declined in importance is that the meaning of the component items can shift rather rapidly. Many sets of items included in accurate and valid scales of intolerance and prejudice during the 1950s and 1960s fail to scale today, and a scale that must constantly be updated doesn't seem to be worth the investment. Finally, social scientists became less convinced that scaling techniques really turned ordinal data into higher levels of measurement. If the superiority of scales to indexes is doubtful, why go to all the added effort?

MEASURES BASED ON OBSERVATIONS

Frequently, social scientists can base their measures on actual observation of the phenomenon of interest. For example, measures based on observations of actual behavior are typical in field research. Researchers have observed and recorded such things as how often students in a particular class interrupt the teacher, how many bicyclists passing an intersection give appropriate hand signals, or how many people give money to a young woman who is displaying a sign saying she is homeless.

Observation yields many measurements in other modes of research as well. Survey interviewers have been asked to notice and report on the quality and cleanliness of respondents' homes, their attitudes toward being interviewed, and their physical attractiveness. Similarly, in most experiments, the researcher has the opportunity to actually observe the subject's reaction to some stimulus, and this behavior can serve as an indicator measuring some concept. Thus, in Chapter 10, we recount an experiment in which the dependent variable involved people leaving a laboratory booth in search of help.

RATES AND OTHER AGGREGATE MEASURES

One of the most obvious problems in using aggregate units of analysis, such as towns, high schools, states, or nations, is that we can't question our cases—a city won't tell you how many robberies it had last year. We have to obtain information on aggregate units in other ways.

Many useful measures are created by summing, or aggregating, data on individuals (or other units) within the larger unit; for example, the population of a city is simply the sum of the individuals living within its boundaries, and the number of robberies is simply the number of times individuals were robbed. Usually, these "raw" numbers are turned into rates.

A **rate** is a proportion or ratio and usually is created by dividing one variable by another variable. The purpose of rates is to create a common basis for comparison across aggregate units so as to allow for differences in such factors as size.

Suppose we wanted to study infant mortality defined as the number of infant deaths during the first year of life. The first thing we would discover is that the United States has far more infant deaths than does Canada and that Canada has far more than Haiti. But that's neither interesting nor surprising because the United States has a population of more than 250 million, while Canada's population is just under 27 million and Haiti has barely more than 6 million people. To see where an infant is more likely to die before his or her first birthday, we need to remove the effect of population. We do this by creating an infant mortality rate: the number of infant deaths per year, divided by the number of live births per year and multiplied by 1,000. The result is the number of infant deaths per 1,000 live births. This, in effect, makes Haiti, Canada, and the United States the same size. And now we find that Haiti's infant mortality rate of 105.6 per 1,000 live births is many times higher than the rate for the United States (10.3), but that the U.S. rate is higher than that of Canada (7.2). You will learn more about rates in the workbook.

There are many indexes used in studies based on aggregate data, including measures of income inequality, segregation, and social mobility. We shall consider several of these in detail in Chapter 8.

Another common technique when using a unit of analysis other than the individual is to use "judges" to rank the cases. **Laurence Iannaccone** (1994) proposed to measure religious denominations on a dimension he identified as their degree of "strictness," the extent to which the group was distinguished from the general American society on the basis of more demanding moral requirements. To facilitate this transformation, he asked 16 experts on American religion to rate each of 16 major denominations in terms of strictness, which he operationalized for them as follows:

> Does the denomination emphasize maintaining a separate and distinctive life style or morality in personal and family life, in such areas as dress, diet, drinking, entertainment, uses of time, marriage, sex, child-rearing, and the like? Or does it affirm the current American mainline life style in these respects?

Each expert was asked to rate each denomination on a seven-point scale as shown below:

Maximal affirmation   Maximal distinctiveness
of American life style   from American life style

1   2   3   4   5   6   7

Iannaccone then added up the total score given a denomination by the experts and divided by 16 (the number of experts) to obtain an average rating score for each denomination. He used these scores to place each denomination on a continuum representing the concept of strictness. To justify this quantification of strictness, Iannaccone reported an extraordinary level of agreement among the experts (correlations of .85 and above). One expert failed to rate one of the denominations, so the score for that group was based on the 15 experts who rated it.

In Chapter 8, we'll examine measures such as these in greater detail.

CODING CONTENT TO GET MEASURES

Sometimes the units of analysis used by social scientists are not humans, or even aggregates of humans, but artifacts created by humans, such as newspaper stories, movies, books, paintings, and so on. While some characteristics can be directly measured, such as number of words in an article, most characteristics require a coder to make judgments about how the case should be categorized. Most research of this type uses *content analysis* to measure variables.

Content analysis is a research technique used to systematically transform nonquantified verbal, visual, or textual material into quantitative data to which standard statistical analysis techniques may be applied.

Content analysis has been applied to books, diaries, speeches, stage plays, newspaper and magazine articles, song lyrics, television and radio programs, movies, ads, and even graffiti.

For example, **Amy Binder** (1993) became interested in the media coverage of heavy metal and rap music after congressional hearings on the need for warning labels had been held. She selected a sample of magazines and found a total of 118 articles that discussed the lyrics of either heavy metal or rap. She found that the articles had either of two basic themes: The lyrics are harmful or the lyrics are not harmful. This could be coded as 1 for harmful and 2 for not harmful. She also coded the justification presented for the position taken. She found that heavy metal and rap lyrics were attacked for different reasons and they also were defended using different rationales. Realizing that these differences might be based on actual differences in the lyrics, Binder coded lyrics of a sample of 10 heavy metal songs and 10 rap songs:

Content of Lyrics	Heavy Metal	Rap
Hard swear words	2	9
Sex, graphic	1	7
Violence or murder of police	0	2
Rebellion against teachers/parents	2	0
Degradation of and violence to women	3	6
Sex, indirect references (innuendo, double entendre)	2	1
Grisly murder, violence, torture	1	0
Sex, group	1	0
Drugs and/or alcohol	1	0
Incest	1	0
Prejudicial slurs	1	1
Suicide	1	0

These results suggested that the differential response of the media to the two types of music was probably justified.

You will learn much more about content analysis in Chapter 11. Similar techniques are used in coding field notes in field research. In such research, the researcher observes individual or groups in "natural" settings, taking copious notes on his or her observations. These notes must later be organized in some fashion. Changing these notes into variables is discussed in Chapter 9.

CONCLUSION

The purpose of this chapter was to introduce the major issues and techniques involved in measuring social scientific concepts. Having begun with variables measuring attendance at stock car races and appreciation of jazz, the chapter came full circle to end with content analysis of heavy metal and rap lyrics. Thus do we confront the real motto of social science: *Practically anything people do, think, or create can be measured and turned into numbers.* And the results of that motto can be seen in the wonderful diversity and richness of social research. Subsequent chapters display more of these diverse accomplishments and will explain the techniques that make them possible.

REVIEW GLOSSARY

• **Constants** are characteristics or aspects of the things being studied that do not vary, but that take the same value. All cities have crimes; therefore, having crimes is a constant, just as breathing is a characteristic of all living human beings.

- **Variables** are characteristics or aspects that take different values among the things being studied and thus can be said to vary. Homicide rates vary across the 50 states, and therefore the homicide rate is a variable as are all aspects of states on which they differ including the proportions of their populations who bowl, eat pizza, or drive pickup trucks. At the individual level, all ways in which people differ are variables: height, weight, religion, hobbies, voting preferences, and the like.

- **Nominal** or **categorical** variables sort cases into categories. Gender is a nominal variable; cases can be separated into one of two categories: male or female. However, the categories of a nominal variable lack intrinsic *order*.

- The categories of **ordinal** variables can be ordered. Thus, for example, people can be ranked along a continuum from liking jazz very much to disliking it very much.

- **Interval** variables are a more precise form of ordinal variable in that the gaps or intervals between categories are of equal quantity.

- **Ratio** variables have meaningful zero points. Thus, we can say that a person weighing 220 pounds is twice as heavy as a person weighing 110 or that someone age 40 is twice as old as someone age 20.

- Social scientists base their research on a variety of **units of analysis**—the "things" a hypothesis directs us to observe. Often these things are individual human beings which are referred to as *individual* units. But they also consist of larger units which are referred to as *aggregate* units.

- To aggregate is to put together; hence, **aggregate data** refer to data based on units of analysis larger than the individual. Because, in a sense, aggregate data describe the social reality surrounding the individual, such data also often are referred to as **ecological data**.

- When we assume that findings based on aggregate or ecological data apply to individuals, we are committing the **ecological fallacy**.

- A variable is **reliable** if it is consistent—if repeated observations give similar results.

- **Inter-rater reliability** assumes that a variable is well measured if it is created or scored by independent raters or coders who achieve a high level of agreement.

- **Test-retest reliability** involves measuring the same cases at two or more times and comparing the results.

- The **alternate forms** approach assesses reliability by comparing two or more independent measures of the same variable.

- The **split halves** technique separates measures collected simultaneously and compares them.

- The **internal consistency** method assesses the reliability of a set of measures of the same concept by comparing all possible combinations of these items and calculating **Cronbach's alpha**, which is the average inter-item correlation for a set of items.

- A variable is **valid** if it actually measures the concept it is meant to measure.

- **Face validity** is the most common basis for establishing validity—to conclude that *on the face of it* the variable obviously measures the concept.

- A somewhat more stringent test of validity is based on the principle that valid measures of the same concept must be correlated. This is called **convergent validity** because the indicators converge on a single, underlying, empirical base that represents the concept.

- **Criterion validity** is the most stringent test of validity and is based on comparing a particular indicator with another that is *known to be valid* and which can, therefore, serve as a **criterion**.

- **Construct validity** is based on the match-up between a measure and the assumptions concerning the phenomena the concept is meant to isolate and identify. For example, the Herfindahl index used to measure religious pluralism rests on an elaborate mathematical model from which its computational formula is derived. In such cases, it often is possible to test the validity of a measure by seeing if the measure meets these underlying assumptions.

- **Cross-case comparability** refers to the need that a variable measure the same thing for each case or unit of analysis.

- Combinations of variables based on survey data often are referred to as **indexes** (or **indices**). When social scientists create indexes based on individual-level data (such as survey questions), they typically proceed simply by adding together the values of some set of indicators. Such an index is based on the assumption that *more* measures of the same thing will yield more sensitive measurements.

- **Scales** are measures created by combining variables using an explicit measurement model. Such models typically include the following three elements:
    1. assumptions about the nature of the component questions, or variables
    2. rules for combining the set of questions, or variables, to produce the measure
    3. techniques for assessing the quality of the resulting measure, or scale

- A **rate** is a proportion or ratio and usually is created by dividing one variable by another variable. The purpose of rates is to create a common basis for comparison across aggregate units so as to allow for differences in such factors as size.

- **Content analysis** is a research technique used to systematically transform non-quantified verbal, visual, or textual material into quantitative data to which standard statistical analysis techniques may be applied. For example, content analysis has been applied to books, diaries, speeches, stage plays, newspaper and magazine articles, song lyrics, television and radio programs, movies, ads, and even graffiti.

Censuses and Samples

Suppose social scientists wanted to know how many people in New York City voted in the last election, how many of them read fashion magazines, or how many New Yorkers watch the news on Channel 7. To answer questions such as these, social scientists could do either of two things. They could ask *every* New Yorker. Or they could ask a much smaller group of people selected to represent the entire population.

In this chapter, we will examine the basic features of each of these two approaches, comparing their strengths and weaknesses.

CENSUSES

According to the Bible (Luke 2:1), Joseph and Mary journeyed to Bethlehem because Caesar Augustus, the Emperor of Rome, had ordered that a census be taken of his Empire; for this purpose, everyone was required to return to their ancestral homes.

As defined in dictionaries, a **census** is an official count of the population and the recording of certain information about each person. In social science, the term is used more broadly and refers to instances when *data are collected from all cases or units in the relevant set.* Thus, in addition to referring to gathering information on the entire population of some group or place, the word *census* also applies to studies based on all of the aggregate units making up some set—such as all 50 states, all nations having a population of 15 million or more, all high schools in North America, or all biographies of movie stars published since 1925.

USING CENSUS DATA

The census ordered by Augustus was one of the first censuses known to have been conducted—several earlier ones took place in ancient Egypt. The purpose of these early censuses was to establish how many people and households there were within the territory under the ruler's control and how much tax they could be expected to pay. Through the centuries, these remained the primary reasons for taking a census. Thus, after seizing the English throne following his victory over King Harold in 1066, William the Conqueror ordered a census to report on the number of persons in every household and to list their significant possessions. An angry Anglo-Saxon chronicler noted:

> *He sent his men all over England into every shire and had them find out how many hundred hides[1] there were in the shire, or what land and cattle the king himself had in the country, or what dues he ought to have in twelve months from the shire. Also he had a record made of . . . what or how much everybody*

---

[1] A hide was a unit of land equal to approximately 120 acres, or a square having sides approximately 800 yards long.

had who was occupying land in England, in land or cattle, and how much money it was worth. So very narrowly did he have it investigated, that there was no single hide nor yard of land, nor indeed (it is a shame to relate but it seemed no shame to him) one ox nor one cow nor one pig which was there left out, and not put down in his record: and all those records were brought to him afterwards. (in Hallam, 1986:16)

These records constitute an immense volume known as the *Domesday Book* (pronounced "doomsday"). It still exists and still provides historians with raw materials on medieval England.

When the Constitutional Convention met in Philadelphia in 1787, the need to apportion the House of Representatives as well as the need to fix taxes caused the delegates to write into the Constitution of the United States the requirement that a census be taken every 10 years. Thus, Article I of the Constitution directs that

Representatives and direct Taxes shall be apportioned among the several States which may be included within this Union, according to their respective Numbers. . . . The actual Enumeration shall be made within three years after the first Meeting of the Congress of the United States, and within every subsequent Term of ten Years, in such Manner as they shall by Law direct.

The first census of the United States was conducted in 1790. The enumeration was done by the U.S. marshals and their deputies (about 650 men) who traveled their districts and gathered the information. After debating what information should be included in the census, Congress settled for listing people by age, sex, race, and whether free or slave. Figure 4.1 shows the percentage of the population in slavery based on the results of the 1790 census. There actually were 17 slaves in Vermont in 1790, but that was not a sufficient number to produce a percentage above 0.00. But there were no slaves in Massachusetts or Maine since slavery was illegal in both states. As would be expected, the percentage of the population who were slaves was highest in the South—in South Carolina, nearly half the population consisted of slaves. But notice that there were substantial numbers of slaves in many northern states, too, especially in New York and New Jersey.

As the decades passed, Congress began to expand the questions asked on the census forms. The 1830 census recorded the number of persons who were "blind, deaf," or unable to speak. In 1840, the census added questions on literacy and insanity. In 1850, the census takers began to ask about all deaths in each household during the past year, and the census reports published the numbers of persons who had died of various causes including suicide and murder (see Figure 4.2). By 1860, questions on income, net worth, occupation, and place of birth were added to the census form. Moreover, the census began to gather a great deal of economic and institutional data—from colleges, businesses, churches, prisons, hospitals, and other institutions.

Figure 4.1   Percentage of Population Who Were Slaves, 1790

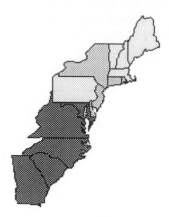

Massachusetts	0.00	Connecticut	1.18	North Carolina	25.55
Vermont	0.00	Rhode Island	1.45	Maryland	32.22
Maine	0.00	New Jersey	6.19	Georgia	35.52
New Hampshire	0.11	New York	6.26	Virginia	39.14
Pennsylvania	0.86	Delaware	15.06	South Carolina	42.95

Today, it is possible to examine microfilms of the actual census forms filled out by the marshals and other census enumerators for all censuses from 1840 through the census of 1930. Seventy years after a census is taken, these forms become available to the public.[2] Consequently, scholars can look up specific individuals and learn much about them, but, more importantly, social scientists can use these data to gain insights into whole communities.

Marion S. Goldman (1981) reconstructed Virginia City, Nevada, during the height of its silver boom and discovered that in 1860, just as the silver rush began, there were only 30 women and 2,206 men in town. By 1870, the sex ratio had shifted to 2,323 women and 4,725 men. Examining information on the occupation and household of each woman, Goldman discovered that a very substantial number of the women who came to Virginia City during this period were prostitutes. Combining these census materials with diaries and old newspaper accounts, Goldman was able to reconstruct how these women fit into the community. The results appeared in her very readable book, *Gold Diggers and Silver Miners: Prostitution and Social Life on the Comstock Lode.*

---

[2] They are located in 12 government census depositories around the nation, and many university and large public libraries have these microfilms as well.

Figure 4.2   Number of Deaths by Various Causes Reported in the Census of 1860

	Males	Females
Accidental	12,399	5,669
Burns and Scalds	1,798	2,477
Drowning	2,660	459
Fall	1,018	303
Firearms	684	46
Freezing	125	14
Lightning	134	58
Railroad	544	55
Other	6,454	2,257
Suicide	794	208
Cutting Throat	57	10
Drowning	40	31
Firearms	109	4
Hanging	249	55
Poison	99	46
Other	240	62
Murder	905	79
Executed	57	4
Alcoholism	1,360	144
Childbirth	—	4,065
Consumption	23,029	25,942
Diarrhea or Dysentery	9,957	8,351
Typhoid Fever	10,321	8,886
Pneumonia	15,804	11,272
Whooping Cough	3,825	4,575

Source: *Preliminary Report on the Eighth Census, 1860*. Washington, DC: U.S. Government Printing Office, 1862.

In similar fashion, **William Sims Bainbridge** (1982) located all of the Shaker utopian communes in the United States censuses of 1840 through 1900 and was able to link members from one census to the next. As a result, he was able to show that the Shakers were exploited as a home for fatherless families in that large numbers of women with children moved into these communities and were provided for, but virtually all of these families left as soon as the children were old enough to work.

But perhaps the most frequent use made of census data, whether old or new, is to aggregate the data to characterize larger units of analysis.

AGGREGATE UNITS

In Figure 4.1, the unit of analysis is not slaves, but states. And the slavery "rate" for each state was created by summing or aggregating the census information for all persons residing in each state.

Census data are a primary source of aggregate data. Nearly all nations, and all of the more modernized nations, regularly conduct relatively detailed censuses and publish the results. Like the United States, some nations conduct a census every 10 years. Others, including Canada, do so every 5 years. Many social scientists use these census materials to create data sets with all nations above a certain size as the units of analysis. Moreover, most nations publish their census data broken down for various geographic units such as states or provinces, counties, cities, neighborhoods, and sometimes even individual city blocks. Social scientists often base aggregate studies on these internal units.

Reynolds Farley and **William H. Frey** (1994) wanted to study trends in the degree of racial segregation in urban America. They based their research on all 232 metropolitan areas and compared measures of dissimilarity for 1980 and 1990. They found that the greatest declines in residential segregation between whites and African Americans occurred in newer, southern and western metropolitan areas.

Stewart E. Tolnay and **E. M. Beck** (1992) wanted to test the hypothesis that lynchings in the South early in the twentieth century caused African Americans to migrate and that this trend, in turn, caused a decline in lynch mob violence as local leaders intervened to prevent a scarcity of cheap workers. To test this hypothesis, Tolnay and Beck assembled data for all counties in 10 southern states for the period 1910–1930. Their data on lynchings did not come from the census—they created this variable from reports in the local press and from many other records and archives. However, the remainder of their variables were from census data. The results supported their hypothesis.

Notice that, in addition to being based on census data, these aggregate units themselves constitute a census. Recall that the definition of a census presented at the start of the chapter identified any set of data based on *all* available units as a census. Thus, since Farley and Frey used all of the U.S. metropolitan areas and Tolnay and Beck used all counties in 10 southern states, their work constituted censuses of metropolitan areas and of southern counties. It doesn't matter that Tolnay and Beck didn't use all counties in the United States; what matters is that they used all counties within the total number they defined as relevant.

POPULATIONS AND "LITTLE" CENSUSES

As noted at the start of the chapter, the word *census* is often misunderstood as limited to data collection operations that include the total population of a nation. But the term applies whenever data are collected from all members of a *population*, no matter how small or large.

As used by social scientists, the word **population** is not limited to human beings but consists of *all units* constituting a set, however that set is defined or delimited. A population also sometimes is referred to as the **universe** of units, in that the word *universe* refers to "all things." "All persons in Denmark" defines a population or universe as does "all children in the fifth grade at Washington School" or "all counties in the United States."

Put another way, it is a census as long as "everybody" is included, whether it is "everybody" in Canada or "everybody" in the smallest village in Mexico. And it also is a census if everybody enrolled in a particular college or even in a particular college class is included. In addition to including "everybody," it also is a census if "every*thing*" is included—all college basketball teams, all chapters of Sigma Chi, or all Pizza Hut restaurants.

The results of a census apply *only* to the set of units on which it is based. A census of Denver cannot be used to describe Colorado, nor can a census of all students enrolled in an introductory chemistry class be used to describe all students in that college or university. However, it is wrong to say that data based on all students in chemistry are *not* a "representative sample" of all students in that school, for the data are *not* a *sample* of *anything*! So long as one is satisfied to describe the population on which a census is based, issues of representativeness cannot be raised. Barring flaws and failures in gathering the data, a census always provides an absolutely accurate picture of the population—if 39 percent of those enrolled in chemistry are females, then that is the exact percentage since everybody was included. Trouble arises when people do not collect data from the population they wish to describe—when they survey all students in a college, for example, rather than all students in Brazil. But a more significant matter arises from the fact that it often is impractical or even undesirable to conduct a census.

SAMPLES

It costs a huge amount of money to conduct a national census. The 2000 census of the United States cost over $3 billion, and the 1996 census of Canada cost $300 million. Only a census will provide detailed information on the smallest geographic units in a nation—for each town and village, for each urban neighborhood or block. Governments often need information on these very small units. Aside from this, however, a census is unnecessary and often is less accurate than are results based on only a *sample* of the population.

A **sample** consists of a set of units or cases *randomly selected* from a population or universe. **Random selection** means that all cases have an equal chance, or at least a known probability, of being included in the sample. The fundamental principle on which sampling rests is: *If all cases have a known probability of being selected for inclusion in a sample, then we can calculate the probability that the group included in the sample is identical to (or representative of) those not included.* Random samples "work" because they are based on the laws of probability, and, therefore, social researchers tend to use the terms **random sample** and **probability sample** interchangeably.

PARAMETERS AND STATISTICS

Samples never are *exactly* like the population from which they were selected. In more technical language, the *statistic* is only an estimate of the *parameter*.

A **parameter** refers to the *true value* of a variable within the population or universe. For example, the actual mean income of a population might be $23,789 per year.

A **statistic** refers to the *observed value* of a variable within the sample. Thus, based on a sample, we might estimate the average income of a population to be $23,186 per year. In this example, the statistic slightly underestimates the population parameter, or the true value. In another sample, the statistic might overestimate the parameter. Another way of putting this is the parameter, being the true value, is fixed, whereas the statistics, being estimates of the parameter, vary from sample to sample.

Except in unusual cases, when the sample includes a very substantial proportion of the population, the degree to which a statistic will approximate a parameter depends entirely on the *size* of the sample. Thus, a sample including 1,000 cases will as accurately estimate population parameters if it is a sample of Canada, of Minneapolis, or of Palm Springs. That these three populations vary immensely in size does not enter into calculating the probability that the statistics approximate the parameters. But a sample of 2,000 will be far more accurate than one of 1,000, and a sample of 25,000 will be extremely accurate.

CONFIDENCE INTERVALS AND LEVELS

The use of samples rather than censuses rests on the fact it is possible to calculate the probability that the statistic falls within a given range around the parameter.

The *range* within which we estimate the statistic to depart from the parameter is known as the **confidence interval**.

You may have noticed how often news stories based on surveys report that the results are accurate, plus or minus some number of percentage points. For example, a story reporting that poll results show one candidate is supported by 57 percent of potential voters may point out that the poll is only accurate by plus or minus 3 percentage points. That means that this candidate may be ahead by as much as 60 percent or by only 54 percent. This is the confidence interval and is calculated on the basis of the size of the sample plus the value of the *confidence level*.

The *probability* that the parameter actually falls within the range stated by the confidence interval is known as the **confidence level**.

When we are dealing in probabilities, nothing is ever 100 percent, dead certain. Thus, our trust in any confidence interval also is less than complete. Pollsters typically report confidence intervals based on a 95 percent confidence level. That is, in the previous example, the odds are 95 percent that the statistic will fall within 3 percentage points, plus or minus, of the actual value, or parameter. Obviously, the higher the confidence level we require, the larger the confidence interval must be. Should we demand a confidence level of 99 percent in our example, the confidence interval may increase to plus or minus 6 percentage points.

A very important application of confidence intervals involves meaningful comparisons. Suppose a poll reported that candidate A was ahead of candidate B by 52 percent to 48 percent. Suppose too that the confidence interval for a sample of the size used by the pollster was plus or minus 3 percentage points. Taking the confidence interval into account, candidate A might be ahead by as much as 55 to 45 percent or *behind* by as much as 49 to 51 percent. The proper conclusion is that these results are too close to call.

SIGNIFICANCE

The basis of research is comparison. When we wish to test a hypothesis, we compare levels of the independent variable to see if they differ in terms of their values on the dependent variable. That is, we determine whether the independent and dependent variables are correlated. Suppose, for example, we wished to test the hypothesis that men are more likely than women to be happy. To test this hypothesis, we could compare the distribution of responses given by males with those given by females to a question about how happy they are. The following results are based on the 1993 General Social Survey—a national sample of U.S. adults.

	Males	Females
Very Happy	30%	33%
Less Happy	70%	67%
	100%	100%
n =	(684)	(917)
p = 0.123		

The first thing this table shows is that the hypothesis is incorrect—in fact, women are slightly more likely than men to say they are very happy. But, before we rush off to tell the world that women are more apt than men to be very happy, we must notice that the comparison reveals only a very small difference. Small differences must raise concern that they reflect nothing but random variation of statistics around parameters.

Whether we can trust differences (or correlations) or should dismiss them as random depends on two factors. First, as with confidence intervals, the probability that a difference is random depends on the *size of the sample*. The difference in happiness between genders probably could not be dismissed as random if it were based on a sample including 25,000 respondents, but it would be dismissed if based on a sample including only 1,000. Second, results are less likely to be random the *larger the difference* or the correlation involved. If 50 percent of women had replied "very happy" as compared with 25 percent of men, it would be very unlikely that this was a random result, given the sample size.

A **test of significance** is a calculation of the odds that a difference or correlation is produced by random fluctuations between the sample and the population, between the parameter and the statistic.

Notice p = 0.123 at the bottom of the table above. This reports the probability that this result is random (based on chi-square, a widely used test of significance). In this instance, the probability that this is a random result is about 1 in 8 (1 divided by 8 = .125). That is, if we tested this hypothesis on 1,000 independently selected samples, we would get a result this large 123 times even if there is in fact no gender difference in the population in terms of happiness. There is no mathematical way to determine just how high the odds of a chance finding must be before we ignore a result. However, through the years, social scientists have settled on the rule of thumb that they will accept no finding if the odds are greater than 1 in 20 that it is random. Thus, social scientists will ignore any difference when probability that the results are random is greater than 0.05. When the probability is less than 0.05, researchers report that their finding is significant beyond the .05 level of significance.

Many social scientists think the .05 level of significance is too lenient and require that the odds of a random result be less than 1 in 100. This is expressed as a result being significant beyond the .01 level of significance. It is not uncommon to obtain odds of 1 in 1,000 that a result is random, and such findings are significant at the .001 level.

Another way of looking at levels of significance is as criteria for accepting or rejecting the null hypothesis (see Chapter 1). Whenever our results fall below the selected level of significance (below .05, for example) we must accept the null hypothesis, which states the absence of a difference or correlation. Our hypothesis above was that men are more apt to be happy than are women. The null hypothesis is that there is no gender difference in happiness. Since the observed difference is not significant, we accept the null hypothesis. Had the difference been significant we would still have rejected the original hypothesis because the gender difference would have been the opposite of what was predicted.

You will gain experience at interpreting significance in workbook exercises.

THE VIRTUES OF SAMPLES

It is true that, other things being equal, a census is always more accurate than a sample. A statistic is merely an attempt to estimate a parameter and a census yields parameters. But other things often are not equal. From a logistical standpoint alone, it is far easier to hire and train interviewers to collect data from a sample including 60,000 people (the number used in the monthly Current Population Survey) than it is to gather data from more than 200 million people. That's why, beginning in 1960, the U.S. Bureau of the Census shifted from using enumerators who went door-to-door interviewing people to using mail questionnaires. But many people who would be willing to talk with an enumerator do not send back their mail questionnaires—some lack sufficient literacy to do so. Thus, nonresponse is a major factor in undercounting, a problem that has plagued the census from the start; President George Washington expressed his conviction that the population was larger than the total of 3.9 million counted by the U.S. marshals in 1790 (Anderson, 1988). Today, the problem of nonresponse is especially important because it is clear that some groups are less well counted than others, which has led many large cities and minority groups to file lawsuits challenging the census results.

In addition to reducing nonresponse, a well-trained interviewer can help people understand questions; hence, more complicated information can be gained with greater accuracy. Moreover, interviewers can make important observations, some of which may encourage more accurate responses. Thus, while the U.S. census will continue to be taken every 10 years,[3] if for no other reason than to provide population statistics for small geographic units as the basis for dispensing federal grants and apportioning political representation, the fact is that the U.S. Bureau of the Census *corrects* the actual census results on the basis of data collected from samples. By focusing on fewer cases, using professional interviewers, and investing in efforts to find everyone who ought to be in the sample, much higher quality data can be obtained than is possible when the focus is on "everybody."

However, sample studies have their flaws too. As will be discussed later in the chapter, nonresponse and other factors often bias the results of surveys. Nevertheless, when confronted with a huge population, researchers usually can obtain better data from a sample than from a census.

SELECTING RANDOM SAMPLES

To select a sample randomly is not simply to select it accidentally. Otherwise, we could select a random sample by standing on a street corner and interviewing whomever walked by. But there is no street corner where everyone, including everyone in the immediate neighborhood, is equally likely to walk by. To select a sample randomly, we must make sure that every member of the population to be sampled has an equal chance of being included, or that everyone has a known probability of being included.

SIMPLE RANDOM SAMPLING

The technique known as **simple random sampling** is based on the principle that all members of the population have an *equal chance* of being selected.

An easy way to select a random sample is the "lottery" technique. The name of every person or other unit included in the population is written on a small disk. All of the disks are placed in a revolving drum and the drum is turned rapidly until the disks are well mixed. Then disks are drawn one by one until they equal the number of "winners" needed to form a sample of the desired size. "Although conceptually simple," as Graham Kalton (1983:9) noted, "this method is cumbersome to execute and it depends on the assumption that the disks have been thoroughly mixed; consequently it is seldom used."

---

[3] It would require a constitutional amendment to dispense with the census.

The usual method for selecting a simple random sample is based on sets of randomly generated numbers. For many years, books of tables of random numbers were published. Recently, however, these books have become obsolete since researchers can easily generate lists of random numbers using their computers. To proceed, a researcher first assigns each case or unit in the population a unique number and then has the computer generate random numbers within the range defined by the lowest and highest number used to number the population. The units or cases whose numbers are included in the random list are thereby in the sample. Often it is possible to use numbers already assigned to the population. Most nations already have assigned numbers to their census tracts and other small geographic units. Most schools already have assigned identification numbers to their students.

SYSTEMATIC RANDOM SAMPLING

A variant on random sampling is known as **systematic random sampling**. It involves sampling a list by selecting the first case randomly and then taking every nth case until the end of the list is reached. First, the researcher must number a list of the entire population and then divide the total number in the population by the number desired for the sample, thus obtaining a **sampling fraction**.

Suppose there are 20,000 units in the population and we want a sample including 1,000 units. This would produce a sampling fraction of 1000/20000, or 1 in 20—that is, every 20th case is the nth case and belongs in the sample. But, to make sure that all cases have an equal chance of being included,[4] the researcher must randomly draw a number between 1 and the nth case—20 in this example— and that will be the first case selected for the sample. If that number were 9, for example, the researcher would begin with the ninth case. Then the researcher would begin counting 20 numbers from the first case selected to what would be the second case—in this example the 29th case (9 plus 20). The count would then begin over, leading to the 49th case (29 plus 20) as the next case to be included in the sample. This process would continue until the entire population had been counted and every 20th case selected.

Statistical discussions of systematic sampling always warn that the method may produce very biased results should the list happen to be ordered on the basis of some cyclical feature having implications for the findings and if the sampling fraction happened to coincide with the length of that cycle. For example, if a list of military personnel is ordered so that every 20th person is an officer, while the other 19 are enlisted personnel, then a sampling fraction of 1/20 would result in a sample that includes either only officers or only enlisted personnel. It seems signif-

---

[4] Technically, while all cases or units in the population have an equal chance of selection, the probabilities of different sets of cases or units being included are not all equal. In this example, the odds of any two cases having numbers lower than 20 both being in the sample is zero. Consequently, only 20 different sets of cases are possible. While this has subtle statistical implications, in practice researchers regard systematic random samples as equivalent to simple random samples.

icant, however, that all of the examples used to illustrate this problem are hypothetical. The chances of encountering such a problem in real life seem remote: first, because it is very unlikely that a sampling fraction would coincide with the length of the cycle and, second, because lists would seldom (if ever) be so precisely cyclical. Nevertheless, when possible to do so, it probably is wise to reorder a list in some relatively unbiased way before numbering it—Kalton (1983) suggested that lists arranged in alphabetical order may reasonably be assumed to be free of any underlying cyclical features.

It is entirely feasible to use simple random or systematic random sampling techniques whenever it is possible to determine the identity and location of each unit in the population. Any group with a membership list is easily sampled in this fashion, for example. Colleges that assign a unique number to each student are easily sampled—one simply begins generating random numbers within the range used by the student number system. Recently, telephone numbers have become a standard basis for random sampling.

TELEPHONE POLLS

If everyone had a telephone and were listed in the phone book, it would be easy to draw random samples of persons in various geographic units—including an entire nation. In fact, more than 97 percent of all homes in the United States and Canada do have telephones. Even though a large number of these phones have unlisted numbers, people with unlisted numbers often receive calls from interviewers conducting surveys. This has prompted a lot of people to wonder how survey interviewers find out their numbers.

As the cost of face-to-face interviews has risen rapidly in recent years, telephone polling has become widespread. Such polls omit households without phones, but researchers have been willing to live with this small bias in exchange for economy and efficiency. Telephone polls are based on two facts: Telephone numbers are unique, and area codes and prefixes (the first three digits in the number) represent specific geographic areas. This means that respondents within the area to be studied can be selected randomly.

Suppose researchers wished to conduct phone interviews with a randomly selected sample of Oklahoma City. First, they would identify the prefixes located within that city. Next, they would use a random number generator to produce a list of phone numbers having these prefixes. The interviewers would then call these numbers and interview randomly selected people in the households of those who answered, eliminating all nonresidential numbers (such as businesses, churches, hospitals, and schools) and all nonworking numbers. If no one answered, the number would be recalled later. Because the numbers were selected randomly, the result would be a random selection of homes having telephones since all numbers, including unlisted numbers, had an equal chance of being selected.

In the early days of telephone polling, a lot of time was wasted on nonworking numbers and on calls to nonresidential numbers. So several companies began to draw random samples of phone numbers for frequently surveyed areas, clean them to eliminate nonresidential and nonworking numbers, and then sell samples of numbers to survey organizations—often selling the same set several times, which is why you often get several calls within a short period. Recently, the telephone companies have taken over this job since their computers can easily and cheaply select random lists of residential phones, thus eliminating the need to sell the same list twice. However, to save money, some polling firms reuse their lists.

If telephone survey researchers have evolved a very efficient method for selecting random samples, they always have had difficulty convincing people to take part in surveys, and this problem has grown acute with the rapid spread of answering machines. Moreover, while telephone polls always have been biased to the degree that homes lacking phones tended to be lower-income homes, answering machines bias results by screening out calls to the higher-income homes (Tuckel and Feinberg, 1991). In effect, answering machines restore the privacy of the unlisted number as people can refuse to answer until they hear who is calling. Further assessments of the accuracy of telephone polls will be included later in this chapter when sources of bias affecting all varieties of survey samples are discussed.

SELECTING STRATIFIED RANDOM SAMPLES

Simple random samples are drawn from the entire population in one step. For a variety of reasons, social researchers often divide a population into several subpopulations or strata, based on information about each unit or case, and then select samples independently from each. This technique is referred to as selecting **stratified random samples**. There are two primary bases for creating strata: the *characteristics* of the units and the *location* of the units.

STRATA BASED ON CHARACTERISTICS

Stratified sampling by characteristics requires that we know the actual proportion of each stratum in the population and that it be possible to draw separate samples from each stratum.

Suppose researchers wanted to sample a population defined as all persons enrolled in a particular university. Besides offering the name and address of each student, the school list included gender and year in school. A simple random sample of this population would produce a statistic for the gender distribution of the students as well as a statistic for the distribution of students as to year in school. These statistics could be compared to the parameter for each of these variables to assess the accuracy of the sample. However, if the researchers divided the student list into subpopulations or strata based on gender and on year in school and then sampled each group separately in proper *proportion*, the statistic would precisely match the parameter on these two variables. That is, there could be no random variation on gender because exactly the correct numbers of males and females were selected. There are many reasons researchers may stratify a population prior to drawing samples, but an important one is that, to the extent that the characteristics used to create strata are correlated with other variables of interest, the sample will yield more accurate results. That is, by reducing the random error on these variables to zero, researchers also reduce the random error in highly correlated variables. For example, among university students, age is highly correlated with year in school. When there is no error in selecting students according to year in school, the random error in their ages also is reduced. Hence, researchers often prefer to select stratified samples.

Stratified samples also often are used in aggregate studies. However, rather than draw independent samples from each of several strata, aggregate studies may sample only one stratum. For example, studies of racial segregation typically begin by stratifying U.S. cities on the basis of the size of their African-American population—usually all cities having more than 100,000 African Americans and those having less (Taeuber, 1983). Then the researchers select samples of blocks from each city in the over-100,000 stratum. An index of dissimilarity is calculated for each block, and the indexes are summed to yield a score for each city. Cities then become the units of analysis.

OVERSAMPLING STRATA

A second major reason to stratify a population prior to sampling involves characteristics that are very unequally distributed across the cases or units making up the population. In these instances, researchers often **oversample** the smaller strata; that is, they often select more cases from a small stratum than its true population proportion—a practice sometimes referred to as **disproportionate stratification**.

Suppose you wanted to study registered nurses and were particularly interested in comparing the careers and earnings of males and females in this occupation. Drawing a simple random sample of 1,000 cases from a national registry of all nurses in the United States would result in about 935 females and 65 males (these are the parameters based on 1999 information). That would be too few males to produce trustworthy statistics. But obtaining 1,000 male nurses would require a sample of about 17,000 cases. This would be extremely expensive and quite unnecessary if it is possible to stratify the sample on the basis of gender and to select independent samples of each gender. By using different sampling fractions, you could randomly select 1,000 female nurses and 1,000 male nurses. Now comparisons between the two genders would be based on equally trustworthy statistics.

Oversampling is quite common in survey studies. When Statistics Canada conducts its annual Canadian General Social Survey, each province is treated as a separate stratum and the smaller provinces are substantially oversampled. This makes it possible to base meaningful comparisons on the 10 provinces. In similar fashion, studies paying special attention to race and ethnicity often substantially oversample minority populations.

So long as researchers are analyzing strata separately and simply comparing the results, no statistical problems arise. However, should the researchers wish to characterize the entire population from which the separate strata samples were drawn, they must reduce the contribution of the oversampled strata until it is proportionate. This will be discussed later in the section on weighting.

CLUSTER SAMPLING

No one has a list of persons enrolled in all of the colleges and universities in Canada or the United States. But it is relatively easy to get a list of all colleges and universities in both countries. Since all students are enrolled in a school, sampling procedures can begin with those "clusters" of students represented by each school. To proceed, the researchers first draw a random sample of clusters—of all colleges and universities, for example. Then a list of all students enrolled in each school is obtained. Finally, a random sample of students is selected from each school.

Cluster sampling is a two-step process in which aggregates or groups of individuals (clusters) are sampled, and then samples of individuals are selected from within each aggregate (cluster). Cluster sampling is commonly used to sample institutional clusters, such as schools, churches, corporations, and service clubs. But the primary use is to draw samples of general populations on the basis of *residential* clusters. As we will see in greater detail next, the way general populations are sampled is based on random selections of blocks (a different method is used in rural areas). To sample Dallas, for example, first a random sample would be drawn from the population or universe defined as all blocks in the city of Dallas. Next, people would be sent to each block selected into the sample to locate all residences and to list all persons residing in each. Then a random sample (or a stratified random sample, in some instances) would be drawn from each block.

PROBABILITY PROPORTIONAL TO SIZE (PPS)

Implicit in the previous discussion is that clusters are all of the same size—that all colleges have the same number of students and that all blocks have the same number of residents. This is obviously not so. Rarely are clusters of the same size—infantry squads and professional sports teams being obvious exceptions. So, if we begin with a random sample of units of unequal size and then select a random sample of equal size from each, the result will be to greatly underrepresent cases from the larger units. There are several solutions to this. One is to assign different weights to persons selected from units of different size in order to restore proportionality—as will be explained in detail later in the chapter. The more common solution to this problem is to draw from each cluster a sample *proportional to the size of the cluster*. That is, a sampling fraction is calculated for each cluster based on the size of the cluster relative to the size of the population. This is called sampling based on **probability proportional to size (PPS)** because it gives each case in each cluster the same probability of being included since cases in large clusters now have as good a chance of being chosen as do those in small clusters. When there is a very great range in cluster sizes, researchers often stratify the sample on the basis of size before selecting the clusters. In these cases, different probabilities of selection must be assigned each stratum on the basis of the proportional size of each stratum in the population.

NORC National Samples

The best national surveys are based on PPS cluster samples. To understand how these studies are really done, let's explore the procedures used by the National Opinion Research Center (NORC) at the University of Chicago—one of the premier survey research organizations in the world.

The NORC national sample uses a variation of the stratified PPS method. First, the nation is divided into clusters, which NORC sampling experts call primary sampling units (PSUs). PSUs are based on counties; the total number of households in each county, as reported by the most recent census, is taken into account. Each metropolitan county is regarded as a PSU, as are many nonmetropolitan counties. Sparsely populated rural counties are merged into adjacent county units to obtain a minimum of 2,000 housing units. Following these procedures, the 3,141 counties of the United States were transformed into 2,489 PSUs to serve as the basis for the NORC national sample to be used during the 1990s.

Because NORC uses systematic random sampling procedures to select PSUs, the second step is to calculate a sampling interval. Since NORC samples are based on 100 PSUs, and since the total number of housing units in 1990 equaled 102,263,678, this total was divided by 100 to yield a sampling interval of 1,022,637. However, this sampling interval was not applied to the full set of PSUs.

Twelve PSUs had a total number of housing units exceeding the sampling interval. Hence, it would have been impossible for them *not* to be included in the sample. These very large PSUs "are sometimes referred to as self-representing; they are so populous that they encompass an entire portion, or zone, of the list by themselves and, in effect, represent no other PSUs" (Tourangeau, Johnson, Qian, and Shin, 1993:7). The 12 were New York, Los Angeles, Chicago, San Francisco/Oakland/San Jose, Philadelphia, Detroit, Dallas-Ft. Worth, Washington, DC, Houston, Boston, Atlanta, and Tampa/St. Petersburg/Clearwater. So, after placing these 12 PSUs into the sample, NORC calculated a new sampling interval based on the target of an additional 88 PSUs and the remaining total of households. Now 7 additional PSUs exceeded the sampling interval, and they too were automatically placed in the sample. These were St. Louis, Minneapolis-St. Paul, Phoenix, San Diego, Baltimore, Pittsburgh, and Seattle. Again, a sampling interval was calculated to select the additional 81 cases. None of the remaining 2,470 PSUs exceeded the new sampling interval of 805,741.

Then each of the 2,470 PSUs was assigned a number range based on its total number of housing units. Thus, the first case might have been assigned a range of 1 through 54,768, the next a range from 54,769 through 589,799, and so on until the last PSU was assigned a range. Then a random number was selected between 1 and 805,742, and the first PSU to be selected was the one within whose range this random number fell. All other PSUs were selected by adding the sample interval to the initial random number on a cumulative basis.

Having selected 100 PSUs, it was time for the second stage: selecting the block "segments." These consist of linked groups of one or more geographically contiguous census blocks. A total of 384 segments was selected from the 100 PSUs—the number of segments selected from each was determined by PPS methods on the basis of the total number of housing units in the PPS relative to the national total.

Not only will NORC use the same PSUs for all of its samples during the 1990s, but the same segments will be used as well. To prevent repeated interviewing of the same households, new samples are drawn within segments for each new study. To select a specific sample, NORC sampling experts divide the total size of the sample by the total number of segments (384), and that yields the number of persons to be included in the sample from each segment. Thus, a sample size of 2,300 would require that six persons be randomly selected and interviewed within each segment.

The reasons for having a sample that is very geographically stable are practical, not statistical. If NORC researchers drew a new sample of PSUs and of segments for each survey, they would have to recruit interviewers willing to travel, and the high costs of their transportation and living expenses would have to be added to the already very high costs of survey research. So organizations such as NORC recruit and train interviewers who live close to the various segments, and this requires that segments stay put.

Figure 4.3 maps the NORC PSUs. Keep in mind that some of these PSUs have many segments and these are scattered randomly within these large metropolitan areas—many of New York's 26 segments are in New Jersey and one is in Connecticut.

Figure 4.3   Map of United States showing each NORC PSU

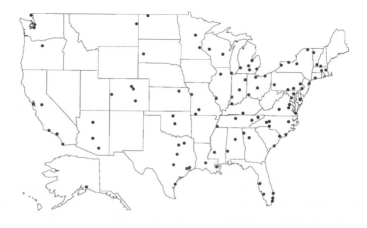

SAMPLING ELUSIVE SUBGROUPS

Social scientists frequently are interested in only a certain subgroup in a population, and often it is extremely difficult to locate their stratum on the basis of available information. For example, no list exists of blue-eyed persons. One way to obtain a sample of the blue-eyed population would be to first select a sample of the entire population, identifying those with blue eyes, and to then interview them. Of course, this would be a far more efficient approach in Norway, where most people have blue eyes, than in Greece, where blue eyes are much less common. *Scarcity* is a major problem in obtaining samples of many interesting subgroups. An additional problem exists for subgroups who simply are *difficult to locate*. How could we obtain a sample of men who have deserted their families? Or of runaway teenagers?

These are precisely the sorts of questions that seem to delight professional sampling experts. The two examples that follow display several of these experts doing their very best.

HOMELESS IN CHICAGO

Peter H. Rossi (1989) obtained funding to study the homeless in Chicago—how many there were, who they were, and how they had ended up homeless. At this time, there were many "expert" and "informed" estimates of the homeless population—both for the nation as a whole and for major cities. The homeless population of Chicago had been estimated at 25,000, and many press reports accepted claims that there were 2 to 3 million homeless in America. It was recognized, of course, that the actual number of homeless was unknown—indeed, New York's Governor Mario Cuomo (1983) argued that the number of homeless never could be known, for this would involve "counting the uncountable."

Rossi disagreed. Based on his long experience, he was certain that the homeless could be counted, and he proposed to begin by counting them in Chicago. Rossi also realized that there was no need for a census of the homeless if it were possible to secure a proper random sample of homeless people. But how do you find homeless people? Since Rossi was a former director of NORC, he naturally consulted Martin Frankel, NORC's staff sampling expert. Together, they formulated a sampling plan.

To begin, they consulted many people who were involved in efforts to help the homeless. The first thing they learned was that homelessness is somewhat seasonal—higher in summer than in winter. So they decided to do their study in two phases—one during warm weather and one during cold weather. Second, they were advised to pay people to be interviewed in order to minimize refusals. The third thing they learned was that the homeless are very mobile during the day, but settle down at night to sleep. So they decided to do their interviewing at night. In addition, they learned that, while few homeless people would pose a danger to interviewers, the night streets in many parts of Chicago—and especially where the homeless congregate—are dangerous. Off-duty police officers were recruited to accompany all interviewing teams.

But Rossi and Frankel still had not solved the basic question of how to draw a sample. Their solution lay in sampling *places*. The homeless, like everyone else in Chicago, are always *someplace*.

At night, some of them are in shelters operated by various charitable groups. So Rossi and Frankel assembled a complete list of shelters and their average number of occupants. From this list, they drew a proportional sample (PPS) of shelters. Members of the homeless population who are not staying in shelters are to be found in a variety of places—in abandoned buildings, in parks, on sidewalks, in doorways, in alleys, in vacant lots, and elsewhere. To sample these people, a complete list of Chicago's blocks was assembled—19,400 in all. Then community relations officers in each of the Chicago Police Department's precincts were asked to rank each block in their precinct in terms of the probability of finding a homeless person there during "the dead of night" (Rossi, 1989:227). The community relations officers consulted with officers on each beat in the precinct to rate each block as apt to have zero or up to five homeless. Blocks then were stratified into those having low, medium, and high densities of the homeless. High-density blocks were oversampled, and low-density blocks were undersampled.

Having figured out where to find the homeless, Rossi still had to define who qualified as a homeless person. After considering many issues, Rossi (1989:12) noted that

> *any current member of a household who occupies a dwelling rented or owned by someone in that household can be said to have a home. Any person who does not own or rent a dwelling and is not a regular member of a household that does so is homeless.*

Rossi's definition of a household was sufficiently broad so as to include hospitals, jails, prisons, and nursing homes.

On the night of the warm-weather phase of the survey, NORC interviewers descended on the shelters in the sample, randomly selected respondents from among those present, and interviewed them. But it was much more difficult for the interviewer teams assigned to the sample of blocks. Beginning at 1 A.M. and continuing until 6 A.M., the interview teams searched each assigned block for homeless people. Each block was completely searched by two different interviewer teams during the night. As Rossi (1989:228) described it:

> *The teams were instructed to walk around each block, go down any alleys and passages, and enter each structure that allowed public access (that did not have a locked door or gate) and to query each person they encountered. Interviewers were specifically instructed to look in parked cars or trucks, to check all-night movies, restaurants, and bars or other places open to the public, and to enter every structure on the block, proceeding as far into each as possible, including searching open basements, roofs, hallways, and any other place they could gain access to without destroying property or taking undue risks.*

The interviewers were instructed to interview each person they encountered using a short "screening" schedule. Everyone they met walking, sitting, or lying down, whether inside a structure or outside, in a car, van, or truck. . . . Sleeping persons were to be gently awakened and then asked for an interview. In all cases the interviewing teams were to announce their purpose, assure the person approached that no harm was intended, and identify themselves and their [police] escorts.

The screening interview was to determine whether the person encountered was homeless. Each was offered $1 to respond to the screening [questions]. If the person was determined to be homeless . . . the interviewer proceeded with a longer interview, offering an additional $4 payment. The longer interview schedule was identical to the one used in the shelter.

On blocks where there were too many people for the interviewing team to handle, all of the homeless were counted and random procedures then were used to determine whom to interview. Later in the year, when the weather was cold, the entire operation was repeated. Amazingly high rates of participation were gained by the survey teams. In the shelters, more than 80 percent of occupants gave completed interviews. Out on the streets, fewer than 20 percent refused to answer the screening questions, and only 4 percent of those screened refused to complete the interview (O'Brien, Loevy, and Roden, 1986).

As a result of this clever sample design, Rossi obtained a very accurate count of the "uncountable." Using various weighting techniques (see the later section on weighting), he was able to project from his sample to the entire population of homeless. His results showed there were an estimated 2,344 homeless people in Chicago during the warm-weather survey and approximately 2,020 during the cold-weather survey. Using information on the length of time people were homeless, Rossi estimated that about 6,000 different individuals would have been homeless in Chicago at *some time* during a year.

Many similar studies have been done since. The results reveal that most estimates of the homeless population were much too high. Distilling many of these studies, Christopher Jencks[5] (1994) estimated that the homeless population peaked at about 400,000 in 1987–88 and has since declined to around 300,000.

AMERICAN JEWS

Over the past 30 years, a number of attempts have been made to select national and local samples of Jews. Some of these studies were based on lists of persons who were affiliated with local temples or with Jewish cultural and charitable organizations. But these studies missed a lot of people of Jewish descent who were not active in Jewish affairs. In response to this problem, rather sophisticated methods

---

[5] Should you wish to know more about who is homeless and why, you may wish to consult Jencks's book.

were developed to identify Jewish names, and samples of persons having such names were selected from the telephone book. Names obtained in this way were then added to those sampled from lists of active Jews. But, even so, many people who should have been included were not. Many Jews no longer have identifiably Jewish names. Given high intermarriage rates, Jewish women married to non-Jews would be missed as would their descendants.

Then, in 1988, the Council of Jewish Federations agreed to sponsor a national survey of the Jewish population to be conducted in 1990. The aim was to have a sample of at least 2,500 Jews and to achieve a random sample of the actual Jewish population.

Barry A. Kosmin of the Graduate School of the City University of New York was assigned the task of locating a comprehensive sample of American Jews. Kosmin consulted **Dale Kulp**, president of the Marketing Systems Group and a sampling statistician. Together, they solved the problems of locating Jews.

All along, it had been obvious that if a researcher selected a large enough sample, he or she would end up with a sufficient number of Jewish respondents. But interviewing 100,000 respondents in order to obtain 2,000 or 3,000 Jews would be prohibitively expensive. However, Kosmin and Kulp recognized an inexpensive way to accomplish the same thing. The ICR Survey Research Group in Media, Pennsylvania, conducts two nationwide market surveys every week using telephone polling methods. ICR was commissioned to add four screening questions to each of its polls:

1. What is your religion? [If not Jewish, then ask]

2. Do you or anyone else in your household consider themselves Jewish? [If not, then ask]

3. Were you or anyone else in the household raised Jewish? [If not, then ask]

4. Do you or anyone else in the household have a Jewish parent?

The screening items were asked of 125,813 randomly selected adult Americans. The phone numbers of all households having at least one person who qualified as Jewish, based on the screening questions, were then used by interviewers who conducted the survey of American Jews. Eventually, interviews were completed with 2,441 individuals who also provided information on other members of their households, totaling 6,514 individuals.

The results then were projected to the population with the following results (Kosmin, et al., 1991):

Born Jewish and gives current religious affiliation as Jewish	4,210,000
Converted to Jewish religion	185,000
Born Jewish with no current religious affiliation	1,120,000

Born/raised Jewish, converted to another faith	210,000
Adults of Jewish parentage with another current religion	415,000
Children under 18 of Jewish parentage, being raised in another current faith	700,000

For the first time, the precise number of American Jews (according to various definitions) was known. However, far more important results are forthcoming from this massive study—more than 20 volumes will report on such matters as fertility, intermarriage, conversion, religious practice, and education.

BIAS

Random fluctuations can cause statistics produced by a sample to differ substantially from the true population parameters. But there exist a number of sources of bias that also can greatly distort even the most carefully selected random samples, especially samples based on individuals.

NONRESPONSE BIAS

Very rarely will every person selected in a sample agree to be interviewed or to fill out a questionnaire. For the 1993 General Social Survey, after excluding addresses selected that turned out not to be dwelling units, the sample included 2,016. Of these, 66 were dropped because of "language problems"—since the GSS population is defined as English-speaking persons age 18 and above. Thus, the target population is itself somewhat different from the total U.S. population.

Of the 1,950 persons making up the target sample, 285 refused to give an interview (several of these broke off the interview after it had begun). Another 18 were classified as unavailable as they kept postponing the interview, and 41 people claimed they were too ill to take part. Consequently, the 1993 GSS is based on 1,606 people, or 84.9 percent of the target population.

As survey results go these days, this is an excellent response rate—far higher than the typical telephone poll, which might obtain an interview from only about 40 percent of the households reached (Tuckel and Feinberg, 1991). On the other hand, Statistics Canada often is able to achieve response rates in excess of 75 percent using telephone polling techniques. The difference probably is the result of more effective interviewers and the added motivation to participate in a "serious" study, conducted by a highly respected government agency, as opposed to market research firms.

But, since it is well known that refusal to participate is not random, when there is a substantial nonresponse rate the results are biased in that they underrepresent certain kinds of respondents including younger people, males, residents of large cities, conservatives, the very poor, and the very wealthy (Smith, 1979, 1983; Statistics Canada, 1991). Chapter 7 will discuss techniques for reducing non-responses.

SELECTIVE AVAILABILITY

Not only do people refuse to participate in surveys, but biases also creep in because certain kinds of people are hard to find and difficult to contact even when found. All surveys based on households have too many women and too few men. There are a number of reasons for this. Men are substantially less likely to have permanent addresses. Residents of jails, prisons, and military posts are excluded from survey samples and have overwhelmingly male populations. In addition, men are less apt to be home when an interviewer calls. In his study of the under-representation of males, Tom W. Smith (1979:10), codirector of the General Social Surveys, reported that it

> seems to result from a deep-rooted problem of nonresponse (perhaps exacer-bated by the overwhelming female composition of the interviewing staff). Males are apparently less accessible and cooperative than females. . . .

College students also are underrepresented in surveys because dormitories, sororities, and fraternities are excluded from samples. To overcome this bias, the national study of Jews was conducted in early summer when many students would be living at home.

AREAL BIAS

As you can see in Figure 4.3, there are many parts of the country lacking a NORC primary sampling unit (PSU). Cluster samples such as this assume that important subsets of the population are reasonably dispersed geographically, at least across major regions. If a group is very concentrated in a relatively small area and there is no PSU in that area, the group's members may have little or no chance to be in the sample.

Prior to designing its 1980s sample, there was no NORC PSU in Utah. Since a very large number of American Mormons live in Utah, the seventh largest religious denomination in the country hardly showed up in NORC samples. The 1980s sample design placed a PSU in Salt Lake City and, consequently, an appro-priate number of Mormons was included in the samples throughout the decade. However, once again no city or county in Utah fell into the 1990s sample, and so once again Mormons are undersampled.

Although there are NORC sampling segments in South Florida, none has fallen into the areas where most Cuban Americans live. Consequently, so few Cubans turn up in NORC samples that they do not even rate a specific entry in the variable reporting ethnic ancestry. Similarly, since there is no PSU in Hawaii, the Asian-American population (especially Japanese Americans) also is quite under-sampled.

It isn't only ethnic and religious groups that are highly concentrated, therefore causing areal biases in cluster samples. Some occupations are highly concentrated, too. With no PSU in Wyoming, Utah, Idaho, or the western part of Texas, cattle ranchers and cowboys will have little chance of being sampled. With no PSU in Nevada or in Atlantic City, professional gamblers will be under-sampled.

Problems such as these are unavoidable when sampling very diverse populations having subgroup concentrations, unless simple random sampling procedures are possible, and they are not in most situations. However, there is a method that can help minimize the impact of biases.

WEIGHTING

Earlier in this chapter, we discussed the need to oversample some strata. For example, selecting a sample of registered nurses including 1,000 females and 1,000 males means that males are overrepresented in the total by nearly 18 times. So long as we analyze the data for males and females separately and only compare the results, this doesn't matter. But suppose we also would like to generate some univariate statistics that apply to the population of registered nurses. If gender has *any* effects on *any* of the variables of interest, then simply to merge the two samples would produce misleading results. This problem is easily solved by *weighting*.

Weighting involves assigning different values (or weights) to each case in order to restore proper proportionality to the sample.

We could use a weighting variable to restore gender proportionality to the sample of nurses. If this were a sample of 2,000 randomly selected cases, we would expect to observe 112 males (5.6 percent of 2,000) and 1,888 females (94.4 percent of 2,000). In our stratified sample of 1,000 males and 1,000 females, we have too few females and too many males. If we weight each male by .112 and each female by 1.888, the weighted distribution of sex will be that expected from a simple random sample. That is, each male in the sample is counted as only .112 of a case, while each female is counted as 1.888 cases. Consequently, when we examine the distribution of any variable in this weighted sample—income, for example—males and females make their correct, proportionate contributions to the totals.

All leading statistical analysis programs allow researchers to apply a weighting variable to their analyses. When weights are assigned so that the size of the weighted sample is the same size as the unweighted sample (as in the example of the sample of nurses), this is called *weighting to the sample size*. This is the most common form of weighting and maintains the proper numbers for significance testing. However, samples are sometimes weighted to population size. Statistics Canada, for example, in its General Social Survey assigns weights so that the distributions reflect the distribution in the population, not in the sample.

Confidence intervals and significance tests should not be used with these weights since the number of cases is enormously inflated. For example, if a survey based on 5,000 cases is weighted back to the total population of Canada, the weighted number of cases will equal the total Canadian population and each actual case will be inflated to represent about 5,600 cases. As a result, many results that would not be statistically significant when based on 5,000 cases would be extremely significant if the number of cases actually were over 28 million.

In addition to correcting for oversampled strata, weighting sometimes is used to "correct" samples against biases. For example, since younger people and males are underrepresented in most surveys, Statistics Canada creates weight variables to increase the contribution of younger and male cases until they are proportionate to the population from which the sample was drawn. Similar weights sometimes are assigned to correct for other biases. For example, since NORC interviews only one person from a sampled household, persons in larger households are undersampled. To correct for this, NORC supplies a weight variable that increases the contribution of persons from larger households to the statistics.

RANDOM/QUOTA "SAMPLES"

A number of the leading opinion polls rely on probability proportional to size (PPS) cluster samples until they have selected their block segments. At that point, rather than use random methods to select the individuals to be interviewed, they assign "quotas" defining the kinds of respondents each interviewer will seek out to interview. The quotas are determined proportional to the demographic composition of the segment in terms of certain key variables, usually race, age, and gender. Thus, one interviewer may be asked to go door-to-door in a particular set of blocks until he or she has located and interviewed a person meeting each of the following quotas:

- a white male over 50
- a white female over 65
- an African-American male under 25
- a white male under 25
- a white female 25–35

Another interviewer in a segment having a different composition in terms of these variables would have a different set of quotas. Because the quotas are determined by the distribution of these traits in the population of each segment, the statistics will match the proportions on these variables. And because PPS clusters were the basis for selecting neighborhoods, the cases will be representative in terms of areas. Nevertheless, because the cases were not selected randomly, strictly speaking there is no mathematical basis for estimating how well they reflect the population of the segments. In practice, researchers analyze random/quota samples as if they were fully based on random techniques.

Why do researchers use quota samples? First, because they are much cheaper and because they seem to work very well. They are cheaper because an interviewer need not call back when a randomly selected case is not at home or not available. Long experience comparing the results of quota samples with those of full-probability samples shows a remarkable match-up. In many ways, this is to be expected. Random methods have been followed down to the level of the segment of blocks. Quotas prevent some nonresponse biases that reduce the accuracy of full-probability samples. As mentioned, probability methods result in an undersample of males. When quotas are assigned, this bias is eliminated—albeit that there may be an overselection of cooperative males. In any event, most poll results reported by the Gallup, Roper, and Harris Polls—to name only the major firms—are based on random/quota samples selected in this fashion.

NONPROBABILITY "SAMPLES"

Sampling "works" only if it is based on random selection of cases, for only then can the laws of probability be used to estimate confidence intervals, confidence levels, and significance—with the exception of random/quota samples as noted. When cases are selected from a population by any other means, the result is not a sample. We have no way of knowing whether the results reliably represent the population—and there are many reasons to suppose they do not.

Nevertheless, many journalistic reports and even a substantial amount of research is based on data sets selected by nonrandom procedures. Sampling experts refer to these as nonprobability "samples"—the quotation marks indicating that they are not really samples of anything. To conclude the chapter, we shall examine the two most common forms of nonprobability "samples."

Snowball "Samples." Often, researchers want to study an elusive population but lack the funds to select a random sample. Usually, this is exploratory research and the researchers merely hope to gain some initial insights. For example, **J. Gordon Melton** (1981) wanted to get some idea of who was joining neopagan American religious groups. So he distributed questionnaires to his pagan acquaintances and asked them to refer him to other pagans to whom he might send questionnaires. Melton started with a small number of respondents, but the number "snowballed" as pagans referred him to other pagans.

A **snowball** "sample" is assembled by referral, as persons having the characteristic(s) of interest identify others.

It is impossible to know how closely Melton's findings would match data based on a random sample of the population of American pagans. But it did yield some interesting leads that might justify a proper follow-up. For example, converts to paganism overwhelmingly came from irreligious families.

Although snowball samples can help social scientists to explore an elusive subset, it is very important to remember that the results cannot be taken as more than suggestive, for they can be extremely biased. For example, some rather bitter factional disputes exist among neopagans and, therefore, it is entirely possible that Melton's data apply to only one of these factions—that his respondents were unwilling to refer him to members of an opposition faction.

SLOPS.   Over the past few years, there has been much "research" on American sexual behavior. The results have received a great deal of media attention, since many of the findings have been nothing short of sensational.

Shere Hite (1987) reported that, of women who have been married 5 years or longer, 70 percent have been unfaithful and 84 percent of those married more than 10 years have had an affair. She also claimed that 56 percent of women married for longer than 10 years have had three or more affairs and 15 percent have had more than one extramarital affair going at the same time.

Samuel S. Janus and **Cynthia L. Janus** revealed in *The Janus Report on Sexual Behavior* (1993) that the decline of sexual activity with age is purely a myth. Their data revealed that, among persons over 65, an amazing 69 percent of men and 74 percent of women were having sex at least once a week, while only 11 percent of men over 65 and 22 percent of women over 65 said they rarely had sex!

In contrast, when the 1993 General Social Survey asked a national sample of American adults "Have you ever had sex with someone other than your husband or wife while you were married?" only 13 percent of women said "yes." This was confirmed by the National Health and Social Life Survey (Laumann, Gagnon, Michael, and Michaels, 1994) conducted in 1992 with a national probability sample of 3,432 respondents age 18 through 59. It found that only 15 percent of women reported extramarital affairs. In similar fashion, the GSS results showed that 13 percent of men over 65 were having sex weekly (49 percent had not had sex at all during the past year) and only 5 percent of women that age reported weekly sex (83 percent reported no sex during the year).

So who's right? The first questions to be asked about *any* survey findings are "How large was the sample?" and "How was it selected?" Both the Hite and the Janus studies were based on large numbers of cases (about 4,500 for Hite and about 8,000 for Janus). But neither was a sample of anything! Indeed, it would be rather hard to imagine a more biased selection of people for reports on sexual activity.

Shere Hite distributed questionnaires to American women in a number of ways. She sent questionnaires to various women's clubs and organizations across the nation, asking them to pass them out to their members. She also received many requests for questionnaires from women who had read one of her earlier books or who heard her appeal for respondents on various talk shows. Hite sent each of these volunteers a questionnaire. Hite claims to have sent out 100,000 question-naires. She also claims that 4,500 were returned—which she characterizes as far above average. In fact, had she begun with a sample of some relevant population, such an incredibly low rate of return would have made her results worthless. In this instance, however, *any* response rate is an irrelevancy. Had everyone who received a questionnaire sent it back, this still would be a *self-selected* group of respondents who might be described as a population—but a population lacking any social reality or relevance. All Hite can legitimately claim is that, of 4,500 women who went to considerable trouble to obtain and complete a questionnaire and who knew the author's views concerning a recent sexual revolution, most were promiscuous. These results apply only to this particular set of women.

Samuel and Cynthia Janus also distributed questionnaires in a variety of ways, mostly through friends and students, many of whom were sex therapists who seem to have distributed the questionnaire to their patients and to their friends and acquaintances. Like Hite, they reveal very little about their "sample," but enough to make it clear that it was an accidental collection of volunteers.

Professional survey researchers call studies such as these **SLOPS**[6]—self-selected listener opinion polls. The acronym was coined to identify the absurdity of call-in polls whereby readers or listeners call alternate 800 numbers to register their approval or disapproval of some proposal.

No matter how large the number of respondents who take part in SLOPS, the results have no credibility. One of the largest SLOPS ever done was based on more than 5 million Americans who sent back postcards in 1936 indicating their choice in the upcoming presidential election. The results of this "poll" by *The Literary Digest* showed that the Republican challenger, Alf Landon, would receive over 60 percent of the vote, while the incumbent Democrat, Franklin D. Roosevelt, would lose with less than 40 percent. As it turned out, it was Roosevelt who received more than 60 percent of the vote. In contrast, that same year the Gallup Poll very accurately predicted the election on the basis of interviews with about 1,400 Americans. The basis for these differing results was precisely the same as in the example of conflicting reports about American sexual behavior noted above.

CONCLUSION

This chapter has four primary aims. The first is to *explain* why and how sampling works. The second is to teach you how to *select* simple random and systematic random samples of various clearly identified populations: students at a particular university, members of an organization, residents of a small town. The third is to enable you to *evaluate* research studies based on more complicated sample designs such as PPS cluster sampling. The chapter is not intended as a "do-it-yourself" guide to these methods, because even the most prominent social researchers do not design such samples for themselves. They rely on a small number of sampling experts who have mastered the intricacies involved. Should you become a social researcher, you will rely on these experts, too. Fourth and finally the chapter is

---

[6] The term is attributed to Norman Bradburn, long-time director of the National Opinion Research Center, at the University of Chicago.

meant to *immunize* you against the rising tide of SLOPS. The next time your local TV station or daily paper reports the results of a voluntary call-in poll, or some other collection of respondents selected by nonrandom means, beware of the results. They may be absolutely accurate (accidents do happen!) or utterly wrong—and there is no way to tell.

REVIEW GLOSSARY

- As defined in dictionaries, a **census** is an official count of the population and the recording of certain information about each person. In social science, the term is used more broadly and refers to instances when *data are collected from all cases or units in the relevant set*.

- As used by social scientists, the word **population** is not limited to human beings, but consists of *all units* constituting a set, however that set is defined or delimited. A population also sometimes is referred to as the **universe** of units, in that the word *universe* refers to "all things." "All persons in Denmark" defines a population or universe as does "all children in the fifth grade at Washington School" or "all counties in the United States."

- A **sample** consists of a set of units or cases *randomly selected* from a population or universe. **Random selection** means that all cases have an equal chance, or at least a known probability, of being included in the sample. The fundamental principle on which sampling rests is: *If all cases have a known probability of being selected for inclusion in a sample, then we can calculate the probability that the group included in the sample is identical to (or representative of) those not included.* Random samples "work" because they are based on the laws of probability, and, therefore, social researchers tend to use the terms **random sample** and **probability sample** interchangeably.

- A **parameter** refers to the *true value* of a variable within the population or universe. For example, the actual mean income of a population might be $23,789 per year. A **statistic** refers to the *observed value* of a variable within the sample. Thus, based on a sample, we might estimate the average income of a population to be $23,186 per year. In this example, the statistic slightly underestimates the population parameter, or the true value.

- The *range* within which we estimate the statistic to depart from the parameter is known as the **confidence interval**.

- The *probability* that the parameter actually will fall within the range stated by the confidence interval is known as the **confidence level**.

- A **test of significance** is a calculation of the odds that a difference or correlation is produced by random fluctuations between the sample and the population, between the parameter and the statistic.

- The technique known as **simple random sampling** is based on the principle that all members of the population have an equal chance of being selected.

- A variant on random sampling is known as **systematic random sampling**. It involves sampling a list by selecting the first case randomly and then taking every *n*th case until the end of the list is reached. First, the researcher must number a list of the entire population and then divide the total number in the population by the number desired for the sample, thus obtaining a **sampling fraction**.

- Social researchers often divide a population into several subpopulations or strata, based on information about each unit or case, and then select samples independently from each. This technique is referred to as selecting **stratified random samples**.

- **Stratified sampling by characteristics** requires that we know the actual proportion of each stratum in the population and that it be possible to draw separate samples from each stratum.

- Researchers often **oversample** the smaller strata; that is, they select more cases from a small stratum than its true population proportion—a practice sometimes referred to as **disproportionate stratification**.

- **Cluster sampling** is a two-step process in which aggregates or groups of individuals (clusters) are sampled, and then the samples of individuals are selected from within each aggregate (cluster).

- **Probability proportional to size (PPS)** is used to compensate for clusters having different numbers of members and involves selecting a sample from each cluster based on the relative size of each cluster.

- **Weighting** involves assigning different values (or weights) to each case in order to restore proper proportionality to the sample.

- A **snowball** "sample" is assembled by referral, as persons having the characteristic(s) of interest identify others.

- **SLOPS** are self-selected listener opinion polls.

Causation and Causal Models

O ne of the most common questions asked in everyday life is "What do you suppose caused that to happen?" This basic human desire to know *why*, to discover the causes of things, gave birth to science. In an important sense, science is the search for causes. Thus, when social scientists propose theories to explain some aspects of social life—to say why and how they are connected— they are attempting to identify causal relationships among concepts. In this chapter, we shall examine fundamental principles concerning cause-and-effect in social research.

CAUSES

What do we mean when we say that something is the cause of something else— that lack of supervision causes delinquency, for example? What we are saying is that something makes something else happen or vary.

A **cause** is anything producing a result or an effect, as when one variable produces or results in variation in another variable.

Translated into the language of social scientific hypotheses, *independent* variables are causes and *dependent* variables are what they cause.

Although we have defined *cause*, notice that the heading for this section is plural: Cause*s*. The reason is that things, and especially the things social scientists are interested in, seldom have only one cause. Thus, the task we face is to identify a number of potential causes of some particular phenomenon and then to sort out each as to whether it actually is a cause and, if so, how important it is compared with other causes.

NECESSARY AND SUFFICIENT

A key to sorting out multiple causes lies in distinguishing between necessary causes and sufficient causes.

When an effect *never* occurs in the absence of a particular independent variable, we refer to this as a **necessary cause** or condition.

If teenagers never are delinquent unless they are poorly supervised, then poor supervision is a necessary condition for delinquency to occur—that is, delinquency would never occur among teenagers who are well supervised.

If supervision were a necessary cause of delinquency, the data would look like this hypothetical table:

	Well Supervised	Poorly Supervised
Delinquent	0%	30%
Not Delinquent	100%	70%

Here we see that delinquency is restricted to only those children who are poorly supervised. Notice that this does not say that *all* teenagers who are poorly supervised will be delinquents, but only that no teenagers who are well supervised will be delinquents. In this example, most poorly supervised children are not delinquent. This means that, in and of itself, poor supervision isn't sufficient to cause delinquency.

When an effect *always* occurs when a particular independent variable is present, we refer to this as a **sufficient cause** or condition.

If all poorly supervised teenagers were delinquents, then this variable is sufficient (or enough by itself) to cause delinquency. If poor supervision were a sufficient cause of delinquency, the data would look like this hypothetical table:

	Well Supervised	Poorly Supervised
Delinquent	30%	100%
Not Delinquent	70%	0%

Notice that the definition of *sufficient cause* does not imply that *only* poorly supervised kids will be delinquents. In this example, 30 percent of well-supervised teenagers also are delinquents.

These examples let us see that causes can be *either* necessary *or* sufficient; they need not be both. However, sometimes a cause is *both* and therefore offers a complete explanation of variations in the dependent variable. A rise in temperature is necessary to turn water into steam. A rise in temperature also is sufficient to turn water into steam. Consequently, an increase in temperature is both a necessary and a sufficient condition to turn water into steam.

However, it also is true that a cause can be *neither* necessary *nor* sufficient but still be a real cause. The relationship between supervision and delinquency is an obvious example. In reality, poor supervision is not a necessary cause of delinquency—many well-supervised children become delinquent. Nor is poor supervision a sufficient cause of delinquency because many poorly supervised children do not become delinquent. But criminologists still regard supervision as a very significant cause of delinquency in that poorly supervised kids are far more likely than well-supervised kids to get into trouble.

A failure to understand these distinctions often has caused a great deal of trouble in interpreting the results of social scientific research (Hirschi and Selvin, 1967). Several examples will be instructive.

"LOST" CAUSES

In her much-admired study of the fate of American utopian communities during the nineteenth century, **Rosabeth Moss Kanter** (1972) dismissed the possibility that religion was a cause of group survival. Having admitted that all of the successful communities had a religious basis, Kanter denied religion was important because many of the unsuccessful communities were organized around a common religious mission, too. She defined *success* as having survived for 25 years or more. Separating her 30 utopian communities on the basis of whether or not they were religious and comparing them on the basis of success produces the following table:

	Religious Utopia	Secular Utopia
Lasted 25 years or more	9	0
Did not last 25 years	8	13

Here we see that Kanter only could have rejected the possibility that religion could be a cause of success by failing to distinguish between necessary and sufficient causes. Obviously, religion is not a sufficient condition for success because 8 of the 17 religious utopias did not succeed. But it is an absolutely necessary condition since *only* religious utopias succeeded. In making this error, Kanter dismissed the strongest single independent variable of the many she examined in her book.

A *Report to Congress on Juvenile Delinquency* (1960) noted that many frequently cited causes of delinquency "are not causes in the sense that if they were removed delinquency would decline." Why? Because many teenagers who are from poor neighborhoods or broken homes, who skip school, or who have other family members who have criminal records do not become delinquents. Indeed, it was suggested that these characteristics are merely correlates of delinquency and "correlation is not causation." As we discuss in the next section, correlation indeed does not prove causation, but lack of correlation proves there is no causation. Here the lesson is that simply because variables are not sufficient causes of delinquency does not eliminate them from causal consideration.

Most variables are neither fully necessary nor fully sufficient causes. When we demand that only necessary or sufficient or even both necessary and sufficient causation counts, we relegate most of what we know about the causes of most social phenomena to the realm of "lost" causes.

CRITERIA OF CAUSATION

How can we demonstrate that some variable is the cause of another variable? To demonstrate causation, three tests or criteria must be met. When *any one* of these is not met, no cause-and-effect relationship can exist between variables.

TIME ORDER

The first test can be quickly explained. **Time order** refers to the sequence of variables. As a criterion of causation, it involves a very simple principle: A cause must occur *before* its effect. Put another way, the principle of cause-and-effect makes no sense backwards. Suppose you claim that eating green apples made you sick. That's entirely reasonable, so long as you ate the apples *before* you got sick. If you already were sick when you ate the apples, the apples could not have been the cause of your illness. Or to claim that the divorce of their parents often causes kids to misbehave, it is necessary to show that the children didn't begin their misbehavior until *after* the divorce. Or to argue that loss of friends causes mental illness, it is necessary to show that people lost their friends before becoming mentally ill. The claim of causation is refuted if it turns out that people first became mentally ill and then lost their friends.

CORRELATION

If something is the cause of something else, then it *must* be the case that the cause and effect *vary in unison*. Changes in the cause must produce changes in the proposed effect. When variables vary or change in unison, they are **correlated**. Suppose you turn a knob on the dashboard of your car and your music plays louder or softer, depending on the direction you turn the knob. The position of the knob and the volume of your sound are correlated. But, if you turn the knob and the volume is not changed, you probably are turning the wrong knob, because there is no correlation.

Correlations can be *either* positive or negative. If the higher their incomes, the more TV people watch, this would be a positive correlation—as one variable increases the other increases too. But, if the higher their incomes, the less alcohol people drink, this would be a negative correlation—as one variable increases the other decreases. Either negative or positive correlations can reflect causation; indeed, a hypothesis must specify the sign of the expected correlation.

SPURIOUSNESS

Two variables often appear to have a cause-and-effect relationship when, in fact, they are correlated *only* because each is correlated with some third, unobserved or unnoticed variable. If you went to any grade school in the world and measured the height of each student and then gave each of them a reading comprehension test, you would find a very strong positive correlation between height and test scores: the taller children would have higher reading scores. Cause-and-effect? Hardly. The taller kids are *older* and older kids read better! Variations in age cause variations in both height and reading ability and thus produce a correlation between the two.

Correlations such as this are called **spurious**. They appear to reflect causation, but they don't. Thus, in order to conclude that a cause-and-effect relationship exists, social scientists must make sure that they are not examining a spurious relationship.

Let's look at a second example. Not long ago, the press reported that a researcher had discovered that elderly men apparently can prolong their lives by marrying young women. This finding was based on comparing men age 50–79 who married young women and men the same age who had not. Those who married younger women enjoyed a 13 percent lower death rate than the others. But what *kind* of elderly man would be likely to seek and win a young wife? Probably not those in failing health. Good health caused the correlation between young wives and living longer. Healthier elderly men both married younger women and lived longer.

We shall add to this discussion of spuriousness later on. In addition, you will gain considerable direct experience with spurious relationships in the workbook exercises. Now, however, let's see how cause-and-effect is related to the different positions of variables in a hypothesis.

CAUSAL ASPECTS OF VARIABLES

Hypotheses *connect* variables and often postulate a cause-and-effect relationship among them. For example, the hypothesis that children in one-parent families will tend to score higher on a delinquency questionnaire than will children in two-parent families connects the variables of family structure (taking two values) and delinquency score (which may take many values) and clearly implies that one causes the other. The *position* of a variable in a hypothesis lets us classify variables into four types: independent, dependent, intervening, and antecedent.

INDEPENDENT VARIABLES

An **independent variable** is hypothesized to be the *cause* of something else. Thus, variation in family structures is hypothesized to cause variations in delinquency scores.

DEPENDENT VARIABLES

A **dependent variable** is hypothesized to be the effect *being caused*. That is, variations in delinquency are hypothesized to be caused by variations in family structures—variations in delinquency are *dependent* on variations in family structure.

Independent Variable	Dependent Variable
1 or 2 Parents ⟶	Delinquency Score

ANTECEDENT VARIABLES

An **antecedent variable** is the cause of spurious correlations between other variables. The word *antecedent* is defined as *going before, prior, or preceding*. We call the source of spurious relationships *antecedent variables* because, as causes, these variables must come before their consequences. We can diagram a spurious relationship this way:

The test for spuriousness is based on the fact that a spurious relationship disappears when the antecedent variable producing that relationship is held constant. That is, social researchers eliminate the variation in a suspected antecedent variable to see whether or not the original correlation remains or disappears. This can be done in several ways, but the logic involved is most easily revealed by cross-tabulation.

Imagine a social scientist who wanted to test the hypothesis (based on her hunch) that tall people are more apt to like basketball than are short people. To do so, she could have selected a random sample of American adults and then sent each a brief questionnaire that, among other things, asked each person his or her height and whether he or she liked basketball. Her initial results could have looked like this:

	Tall	Short
Likes Basketball	65%	45%
Doesn't	35%	55%
	100%	100%

These results supported her hypothesis—tall people are more likely (65%) to like basketball than are short people (45%). However, because she was worried about spuriousness, the sociologist might have decided to control gender as a likely antecedent variable. That is, gender is antecedent in time to both height and reactions to basketball, is strongly related to height, and might be related to liking basketball. To test for spuriousness, she could have held gender constant by separating people into two groups, one of males and one of females, and reexamined the original relationship *within* each group. The results might have looked like this:

	Males		Females	
	Tall	Short	Tall	Short
Likes Basketball	85%	85%	25%	25%
Doesn't	15%	15%	75%	75%
	100%	100%	100%	100%

Look at the data for men. The relationship between height and liking basketball has disappeared. Tall men are not more likely than short men to like basketball—85 percent in each group like it. Now look at women. Here, too, the relationship has disappeared. Tall women are no more likely than short women to like basketball—25 percent in each group like basketball. The original relationship between height and liking basketball was spurious.

This illustrates the principle of holding a suspected antecedent variable constant. Gender was not allowed to vary when the social researcher created separate groups of males and females.

INTERVENING VARIABLES

An **intervening variable** is hypothesized to be the *link* between an independent and a dependent variable.

Why did our hypothetical social scientist formulate the hypothesis that children in one-parent families will score higher on a delinquency questionnaire? Because, other things being equal, children will be less *supervised* in one-parent families. That is, the mechanism that links family structure to delinquency is the amount of supervision, and we can expand the hypothesis to a three-step causal chain in which the intervening variable is placed between the independent and dependent variables, where it intervenes, or comes between them, as shown:

If supervision is in fact the link between family structure and delinquency, then something rather interesting ought to turn up in the data. When we compare one- and two-parent families with the *same level* of child supervision, there should be *no difference* in their children's delinquency scores! That is, when an intervening variable is held constant, the relationship between the independent and the dependent variable disappears. The difference between this outcome and spuriousness is that an antecedent variable produces a spurious relationship because it is the cause of both variables in the relationship (as gender causes both height and liking basketball), while an intervening variable is part of a causal chain. The independent variable causes the intervening variable, which, in turn, causes the dependent variable. If that link is broken, then the independent variable is unable to influence the dependent variable.

Parents are not the only potential source of child supervision—sitters, day-care personnel, or relatives (such as grandparents) who share the home can provide close supervision. Consequently, not all one-parent families will provide poor supervision. By the same token, not all two-parent families will provide good supervision. If supervision is the real cause of the differences between children from one- and two-parent families, then the well-supervised children from one-parent homes should not be particularly prone to delinquency, whereas the poorly supervised kids from two-parent families should be prone to delinquency. If this turns out to be the case, then we have confirmed the hypothesis that supervision is the intervening variable.

Here, too, a hypothetical cross-tabulation clarifies the logic involved. First, we separate children on the basis of their degree of supervision. Next, we compare children from one-parent families with those from two-parent families in terms of whether or not they score high on delinquency. If supervision is the intervening variable, then there should be no relationship between family structure and delinquency *within* groups having similar levels of supervision.

	Well Supervised		Poorly Supervised	
	1 parent	2 parent	1 parent	2 parent
Delinquency:				
High	10%	10%	45%	45%
Low	90%	90%	55%	55%
	100%	100%	100%	100%

These results support the hypothesis that supervision is the link between family structure and delinquency.

Keep in mind that antecedent and intervening variables can be distinguished only on the basis of time order—antecedent variables occur prior in time to either of the other variables in a relationship, whereas intervening variables occur after the independent variable and before the dependent variable. There is nothing in the tables that tells us what the time order is. This must be determined in other ways—from the theory, from the research design (see Chapter 6), by collecting additional data, or from common sense (it is self-evident, for example, that gender is determined at conception, long before people become tall or short and long before they know anything about basketball).

Let's look again at the hypothetical three-variable relationship involving gender, height, and liking basketball. When the relationship between height and liking basketball vanished within gender groups, we concluded that it was a spurious relationship. But it is entirely plausible that rather than being spurious, height would continue to influence liking basketball among both males and females even though males are more apt than females to like basketball. That is, taller women and taller men are both more likely to have played basketball in school and thus to have become fans. What we are proposing is that *both* variables will influence liking basketball. If this is true, then the tables will look like this:

	Males		Females	
	Tall	Short	Tall	Short
Likes Basketball	85%	55%	55%	25%
Doesn't	15%	45%	45%	75%
	100%	100%	100%	100%

Looking at the table for men, we see that tall men (85%) are more likely than short men (55%) to like basketball. Among women, tall women (55%) are more apt to like basketball than are short women (25%). Now, compare tall men and tall women—the tall men (85%) are more likely to like basketball than are the tall women (55%). Among short people, men (55%) are more likely than women (25%) to like basketball. The conclusion is that *both* variables influence liking basketball and that these effects are independent of one another.

This is an example of multiple causation, of a dependent variable having two or more causes.

CAUSAL MODELS

In social science, things seldom have a single cause. Consequently, the explanatory power of the field rests on multiple causation—on constructing theories specifying several independent variables and how they influence one another as well as the dependent variable. To test such theories, it is necessary to be able to examine a number of variables simultaneously—that is, to construct what social scientists call *causal models*.

Causal models are statistical descriptions of a specific set of empirical data that attempt to identify and measure all of the relationships among some set of independent variables and the dependent variable. Often, a causal model is designed to gauge dynamic relationships among the variables based on deductions from a theory.

When a causal model includes only three or four variables, it often is possible to use cross-tabulations to create the model. But, as the number of variables increases, the number of combinations of categories of the variables will increase faster. For example, when we were comparing tall and short people as to whether or not they liked basketball, the cross-tabulation required four cells. When we added gender as a third variable, the size of the table doubled to eight cells. Suppose we added a fourth variable having three categories; now the table would have 24 cells. Or suppose we were interested in variables such as age, years of school, and income and each variable took seven values. Each combination of two variables would produce a table having 49 cells, and the three-variable cross-tabulation would result in 343 cells. If a sample of 1,700 persons were absolutely evenly distributed across the cells of this table, there would be only five respondents in any cell. Therefore, as the number of cells increases, the number of cases in each cell will decrease, and soon the summary statistics (such as percentages) become quite unreliable. Consequently, cross-tabular analysis is very limited in the number of variables that can be examined at the same time.

To overcome these limitations, social scientists use various statistical techniques to analyze the effects of many variables at the same time. There are several such techniques, but the one most often used is known as *regression*. The remainder of this chapter introduces you to the basic logic of regression and teaches you how to interpret regression results. Sometimes, dozens of variables are included in one regression analysis, but to understand the logic involved and to learn how to interpret the results, it is sufficient to use only three or four variables at a time. As you will see, regression models are easy to interpret and to compare.

INTERPRETING REGRESSION MODELS

All regression models are based on correlation coefficients. As you know from doing workbook exercises, correlation coefficients tell us the strength of the relationship between two variables. Correlations vary from 0.0 (no relationship) to 1.0 (a perfect relationship) and may be either negative or positive. The trouble is that ordinary correlation coefficients are even more limited than cross-tabulations in that they are limited to examining only two variables at a time.

The statistical technique known as **regression** uses the correlations between each pair of variables in a set of variables to calculate relationships among the entire set.

In cross-tabular analysis, we focus exclusively on the relationships among variables—examining the relationship between two variables while controlling for a third. In regression analysis, we also look at changes in relationships when other variables are controlled, but, in addition, we look at the combined effects of the independent variables on the variation in the dependent variable. This will be clear in the following example.

Suppose we want to understand more about the problem of homelessness. We know the homelessness rate—the number of homeless per 100,000 population—for each of the 50 states. Our initial explorations reveal that the homelessness rate is highly correlated (.652) to rental costs as measured by the median monthly rental—there is more homelessness in states where rents are higher. We also discover that homelessness is higher in states with higher rates of cocaine addiction—the correlation is .606. But suppose we want to examine the impact of each of these independent variables on homelessness at the same time—as we did with height and gender in the previous example. We use regression analysis and obtain the results shown in the graphic display below.

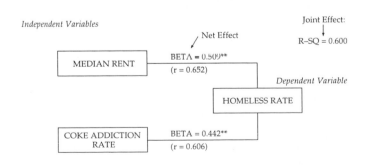

NET AND JOINT EFFECTS

Let's examine the elements shown in this graphic. Beneath each of the horizontal lines is the value of r, which is Pearson's correlation coefficient. These are, of course, the same as reported on the previous page. However, correlations report *only* the relationship between *two* variables. But often the independent variables are correlated with one another as well as with the dependent variable (as they are in the previous example). What regression does is sort out the independent contributions of two or more independent variables.

Notice that above each line in the diagram is the word *BETA*, followed by a numerical value. This stands for the **standardized beta**, which estimates *the independent effect of each independent variable on the dependent variable*. The value of beta is, therefore, the *net* effect of each independent variable on the dependent variable. What we discover in this instance is that both variables help to explain homelessness rates. Each has a net effect—we know this because each beta is significant as indicated by the two asterisks.

Now look at the upper right corner of the graphic and read: R-SQ = 0.600. This stands for R^2, which is *a measure of the* **combined**, *or* **joint, effects** of the two independent variables on the dependent variable. R^2 can be converted to a percentage by moving the decimal point two places to the right. Hence, we can see that rental costs and cocaine addiction together account for 60 percent of the variation in homelessness rates across the states. Put another way, if all states had precisely the same median rent and the same rate of cocaine addiction, there would be 60 percent less variation in their homelessness rates.

To review: Regression reports the *joint* effects of a set of independent variables on the dependent variable (R^2) and the *net* effects of each independent variable (beta). You will learn a bit more about the basis of regression as you use this technique in doing the workbook exercises.

Regression analysis is used with both aggregate and individual data, but the mathematical basis of the technique assumes that variables are at least interval. Fortunately, when ordinal measures are used, the results do not seem to be unduly distorted. Even more fortunate is that researchers have devised techniques for using nominal, or categorical, measures as independent variables. We will discuss these techniques later in the chapter. For now, however, we will use only ordinal or higher measures.

Let's examine a regression analysis using survey data. Suppose we discover there is a strong positive correlation between education and liking Big Band music—music from the 1930s and 1940s recorded by the popular dance bands led by such stars as Benny Goodman, Duke Ellington, Artie Shaw, Tommy Dorsey, and Count Basie and featuring singers such as Frank Sinatra and Ella Fitzgerald. Moreover, there is a strongly positive correlation between liking the Big Bands and age—older people are more apt to like this music. Suppose we want to know which of these independent variables had the greater net effect.

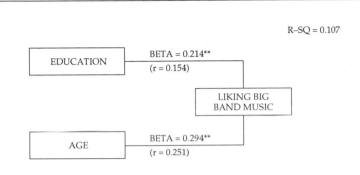

R–SQ = 0.107

Here we see that both variables have a net effect, but age is a bit stronger than education. As for a joint effect, education and age together account for 10.7 percent of the variation in liking Big Band music.

SPURIOUSNESS

Regression analysis also is very useful as a method of testing for spuriousness. Suppose we are interested in explaining variations in the alcohol consumption rate among the 50 states. We discover that the *Playboy* magazine circulation rate is very highly correlated (.622) with the alcohol consumption rate. We also discover that the proportion of male households[1] in each state also is highly, positively correlated with alcohol consumption (.741). Some might argue that reading *Playboy* encourages people to drink. Others might suggest that this is a spurious relationship, that both *Playboy* circulation and drinking will be higher in states having a lot of men living on their own. Here are the regression results:

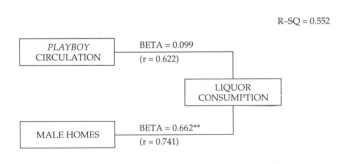

R–SQ = 0.552

---

[1] Households without an adult, female occupant.

Playboy circulation has no significant net effect on liquor consumption when the proportion of male homes is controlled. The beta for *Playboy* is insignificant, and only male homes has a significant beta. That is, the whole effect is caused by the male homes variable. How might we interpret this? Magazines don't drink liquor! People do. What these data suggest is that it probably is the case that men who live alone or with male roommates drink more liquor *and* are more likely to buy *Playboy*. Thus, we have discovered a spurious relationship. Regression calculates net effects by holding all other independent variables constant, so when a variable has no net effect, the original correlation was spurious *if* the variables or variables with net effects are antecedent variables.

Notice that we can say only that it is "probably the case" that men living alone drink more than do other people. To say these data show they do would be to commit the ecological fallacy. These are aggregate data and show only that in states with more male households there is more alcohol consumed. While it seems very likely that these results reflect that men in male households do drink more, to know for sure we would need to examine data based on individuals. In fact, surveys do show that men living alone, especially divorced men, drink more.

SUPPRESSOR VARIABLES

Sometimes when two independent variables are highly correlated with one another and each is strongly correlated with a dependent variable, but in the opposite direction, the correlations will suggest there is no relationship between the independent and the dependent variables. That is, each variable may be suppressing the effects of the other. A **suppressor variable** causes two variables to appear *not to be correlated* when, in fact, they are. In a sense this is the opposite of spuriousness. Here is an example:

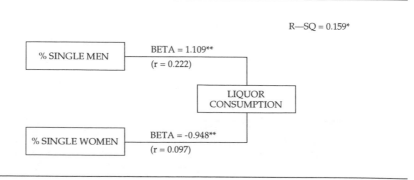

Had we looked only at correlations, we would have concluded that the percentage of single males in the population was only weakly correlated with liquor consumption, while the percentage of single females was unrelated to drinking. But we would have been wrong!

When the effects of these two independent variables are untangled from one another, we find that each is extremely strongly related to liquor consumption, but in opposite directions. That is, a higher percentage of single males stimulates liquor sales, whereas a higher percentage of single females decreases liquor sales. In the two-variable correlations, these opposite effects are canceling one another out, and only regression lets us see the true state of affairs.

We also see that, unlike the correlation coefficient, betas can take a value greater than 1.0 and that in this case the beta between the percent of single males and liquor consumption does.[2]

The graphic presentations of regression we used make it easy to interpret the findings, which is why a similar graphic screen appears in the regression function of MicroCase. However, such graphics take up too much space to be included in social science journals even when they involve only three or four variables and regressions often involve dozens of variables. So let's see how we might report a regression using the suicide rate (suicides per 100,000 population) as the dependent variable.[3]

To see the *joint effects* of these independent variables, find R^2 in the lower left corner of the table on the following page, which indicates that these three variables together account for almost two-thirds of the variation in suicide rates (65.3%). To see the *net effect* of each variable, read down the second column of numbers under the heading "Standardized Betas." The betas shown in the graphics were the standardized betas. Because the significance of an independent variable's effect is based on calculations involving its unstandardized beta, many journals prefer that these values be reported as well as those for the standardized beta.

---

[2] This can occur because a standardized beta tells us that a change of one standard deviation in the independent variable produces a change of how many standard deviations in the dependent variable. It is entirely possible (if infrequent) for a change of one standard deviation in the independent variable to cause a change of more than one standard deviation in the dependent variable, as it did in this example. Standard deviation is covered in all statistics courses.

[3] There is no set, single format for presenting regression results, but this example would meet the requirements of most journal editors.

Regression Analysis of State Suicide Rates
Dependent Variable: Suicide Rate

Independent Variable:	Unstandardized Beta	Standardized Beta	Standard Error	Value of t
Percent all-male households	0.883	0.574	0.175	5.043**
Annual income per capita	−0.001	−0.597	0.000	−6.258**
Church membership rate	−0.087	−0.353	0.027	−3.276**

$R^2 = 0.653$ ** P < .01
n = 50

Standardized betas have the advantage of comparability since the raw values of beta have been "standardized"—that is, converted to a common base. Unstandardized betas are greatly influenced by the metric on which any given variable is based and thus do not allow meaningful comparisons. Notice that, by comparing the standardized betas, you can see that church membership has a smaller net effect on suicide than does either income or male households, but you cannot see that by comparing the unstandardized betas. For now, ignore the standard error. In regression analysis, significance is based on a statistic known as t. All of the betas shown are significant beyond the .01 level as indicated by the two asterisks following each t value. Finally, n indicates the number of cases (or units of analysis) on which the results are based—in this instance, 50 states.

Regression analysis is not limited to three or four variables—regressions involving dozens of independent variables often appear in social science journals. But the interpretation of these models is the same no matter how many variables are included.

COMPARING MODELS

The regression results concerning suicide rates are very complete. Journal articles often omit some of this information. Some articles report only the unstandardized betas while others report only the standardized betas. Some do not report the standard errors or the actual value of t. But all journal articles based on regression analysis report significance for each variable and all of them report the value of R^2. Hence, to compare two or more models, each of which is attempting to explain the same dependent variable, first inspect the independent variables making up each model to see what each includes and which variables contribute a significant net effect. Then compare the R^2 produced by each model to see which one explains more variation in the dependent variable. Let's examine an example.

William Brustein (1991) came to suspect that the accepted wisdom about the electoral success of the Italian Fascist Party in 1921 (led by Benito Mussolini) was wrong. Ever since the Fascists came to power it was believed that voters chose them because voters had become more frightened of the "Red Menace" posed by Socialist and Communist parties. Brustein's initial research suggested that, rather than being driven into the arms of fascism, Italian voters backed the Fascists on the basis of rational self-interest, or, as he put it, "Many individuals chose the Fascists because the Fascist program most closely represented their material interests" (1991:652).

Brustein used as his units of analysis each of Italy's 61 provinces that had at least a third of their populations engaged in agriculture. His dependent variable was the percentage of the vote won by the Italian Fascist Party in 1921. The concept of material interests was operationalized as the mode of production and measured by an index based on several variables including farm size; proportions of the agricultural labor force who were owners, tenants, sharecroppers, or laborers; and average income per farm. The higher a province scored on this index, Brustein wrote, the more "it represents my idea of a district whose mode of production would be most receptive to the Fascist agricultural program" (1991:661). In addition to mode of production, Brustein measured these five variables: (2) proportion urban, (3) proportion of new (first-time) voters, (4) southernness of the province, (5) proportion voting for the ultra-Catholic Popular Party in 1919, and (6) the decrease in the proportion voting Socialist, 1919–21. His initial model did not include the Socialist voting variable. His second model included it. Let's compare the two:

Dependent Variable: Fascist Vote in Italy, 1921

Independent Variables	Model 1		Model 2	
	Unstandardized Beta	t value	Unstandardized Beta	t value
Mode of production	.91	3.17**	.56	2.15*
Proportion urban	−.04	−.26	−.03	−.25
Proportion new voters	−.00	−.01	.01	.19
Southernness	−7.57	−2.48*	−4.02	−1.44
Popular Party vote, 1919	−.29	−2.76**	−.14	−1.47
Decrease in proportion voting Socialist	—	—	.36	4.30***
R^2 =	.42		.57	

n = 61

* $p < .05$ ** $p < .01$ *** $p < .001$

In Brustein's first model, only three variables are significant: mode of production, southernness, and vote for the Popular Party in 1919. Urbanism and new voters contribute no significant net effect. Overall, this model accounts for 42 percent of the variation in the Fascist vote. In the second model, only two variables are significant: mode of production and the decrease in the vote for the Socialists. However, this model explains 57 percent of the variation in Fascist voting. Consequently, the second model is the better one. Brustein (1991:662) concluded his research paper this way:

> The analysis strongly suggests that the increase in the Fascist vote in 1921 cannot be attributed simply to an influx of new voters who were reacting to the "Red Menace." Rather, it was the result of voters switching from the Socialists to the Fascists. Also, the analysis supports the theory that the people's economic activity and property rights are major determinants of fascism's popular support.
>
> Although the "fear of socialism" undoubtedly was a major determinant of the Italian social elite's decision to support fascism, I feel that its explanatory value has been overstated. What has been neglected in studies of the rise of Italian fascism is the extent to which individual political preferences result from rational calculations of their material interests.

Using Categorical Variables

Mathematically, regression analysis requires that all variables included in a model be of at least the interval level of measurement. We have relaxed this assumption to include ordinal variables. Recall that ordinal variables can be ordered along some dimension such as from high to low or small to large. This requirement prohibits the inclusion of nominal or categorical variables, which consist of qualities, not quantities, such as race and gender. This is a very serious limitation for social scientists working with survey data. For example, anyone wishing to construct an empirical model of support for capital punishment knows that both race and gender are of very great importance. To omit them is to gut one's model. Eventually, researchers found a way to get around this limitation by transforming categorical variables into what are called *dummy* variables.

A **dummy variable** is a categorical variable recoded to assign the values 0 and 1 for *each category* so that 1 indicates the presence of the category and 0 represents its absence.

When a categorical variable has only two categories, as in the case of sex, a dummy variable assigns 0 to one category and 1 to the other. For example, men can receive the value 0 and women can receive the value 1. Now comes a subtle change in the logic of interpreting results. In a cross-tabulation, this variable would represent males and females. But, in a regression, the variable now represents female/nonfemale. The *mean* of this variable now is proportion of females in the sample. When used in a regression model, if the beta for this new dummy variable for gender is positive (and significant), we can interpret that to mean that women score higher on the dependent variable than do men. If the sign is negative, then men score higher than women.

When categorical variables have more than two categories, then more than one dummy variable must be created. National surveys typically code three racial categories: white, black, and other. To transform race into a dummy variable, we recode the categories as follows:

WHITE: transformed into white/nonwhite, with white being coded 1 and all others 0.

BLACK: transformed into black/nonblack, with black being coded 1 and all others 0.

We do not create a dummy variable for other because this group already is represented by the two dummy variables already created since they are the only respondents who are scored as 0 on both dummy variables. If we used a third dummy variable for other/nonother, the regression equation would be *overdetermined*. The rule is this: When creating dummy variables, *never* create more than the total number of categories *minus one*.

Now let's see how these dummy variables work in a model in which support for capital punishment is the dependent variable. This is a three-point ordinal variable in which those in favor of capital punishment are scored highest.

Regression Analysis of Support for Capital Punishment

Dependent Variable: Support for Capital Punishment

Independent Variable:	Unstandardized Beta	Standardized Beta	Standard Error	Value of t
FEMALE/NOT	−0.182	−0.111	0.041	−4.413**
WHITE/NOT	0.277	0.119	0.104	−2.653**
BLACK/NOT	−0.117	−0.043	0.120	−0.969

R^2 = 0.040
Prob = 0.0000

Here we see that women are not as favorable as men toward capital punishment—the dummy variable of female/nonfemale has a significant negative effect. Race also matters, as the dummy variable of white/nonwhite has a significant, positive effect. While the effect for black/nonblack is negative, indicating that blacks are less supportive of capital punishment than are others (mainly whites), the effect is not significant. This is because the white/nonwhite variable already has explained most of the variance caused by racial differences.

You also may notice that the value of R^2 is small but highly significant (p = 0.000). Survey data are far more subject to measurement error than are most aggregate data and consequently produce correlations and betas far smaller than is often the case for aggregate data—as was clear in the examples presented before.

When constructing regression models based on individual level data, it is important to pay close attention to the significance of the model as a whole, because with larger samples quite small values of R^2 will be significant.

DICHOTOMOUS DEPENDENT VARIABLES

Standard regression analysis is inappropriate with dichotomous *dependent* variables. A **dichotomy** is anything that consists of only two values.

Suppose you wish to study smoking and your data are limited to information that a person does or does not smoke. In this circumstance, you would create an empirical model of smoking by using a technique known as *logistic regression*. Although the term *logistic regression* appears on the menu of the student version of the MicroCase Analysis System, it is not among the tasks available in this student version. In general, logistic regression is performed and interpreted in much the same way as regular regression.[4] Fuller treatment of logistic regression awaits you in another course—or your instructor may wish to deal with the topic in lecture.

CONCLUSION

This chapter should give you an adequate intellectual and practical grasp of basic issues concerning causation and of the analysis tools used to assess causal relations among variables. Examples began with the simple cross-tabulation methods used since the earliest days of social research and concluded with the far more powerful tools of empirical model building.

REVIEW GLOSSARY

- A **cause** is anything producing a result or an effect, as when one variable produces or results in variation in another variable.

- When an effect *never* occurs in the absence of a particular independent variable, we refer to this as a **necessary cause** or condition.

- When an effect *always* occurs when a particular independent variable is present, we refer to this as a **sufficient cause** or condition.

- **Time order** refers to the sequence of variables. As a criterion of causation, it involves a very simple principle: A cause must occur *before* its effect. Put another way, the principle of cause-and-effect makes no sense backwards.

---

[4] Regular regression often is referred to as OLS regression, or ordinary least squares regression.

- If something is the cause of something else, then it *must* be the case that the cause and effect *vary in unison*. Changes in the cause must produce changes in the proposed effect. When variables vary or change in unison, they are **correlated**.

- Two variables often appear to have a cause-and-effect relationship when, in fact, they are correlated *only* because each is correlated with some third, unobserved or unnoticed variable. Correlations such as this are called **spurious**. They appear to reflect causation, but they don't.

- An **independent variable** is hypothesized to be the *cause* of something else. Thus, variation in family structures is hypothesized to cause variations in delinquency scores.

- A **dependent variable** is hypothesized to be the effect *being caused*. That is, variations in delinquency are hypothesized to be caused by variations in family structures—variations in delinquency are *dependent* on variations in family structure.

- An **antecedent variable** is the cause of spurious correlations between other variables. The word *antecedent* is defined as *going before, prior, or preceding*. We call the source of spurious relationships *antecedent variables* because, as causes, these variables must come before their consequences.

- An **intervening variable** is hypothesized to be the *link* between an independent and a dependent variable.

- **Causal models** are statistical descriptions of a specific set of empirical data that attempt to identify and measure all of the relationships among some set of independent variables and the dependent variable. Often, a causal model is designed to gauge dynamic relationships among the variables based on deductions from a theory.

- The statistical technique known as **regression** uses the correlations between each pair of variables in a set of variables to calculate relationships among the entire set.

- The **standardized beta** estimates *the independent effect of each independent variable on the dependent variable*. The value of beta is, therefore, the *net* effect of each independent variable on the dependent variable.

- R^2 measures the **combined**, or **joint, effects** of the two independent variables on the dependent variable. R^2 can be converted to a percentage by moving the decimal point two places to the right.

- A **suppressor variable** causes two variables to appear *not to be correlated* when in fact they are. In a sense this is the opposite of spuriousness.

- A **dummy variable** is a categorical variable recoded to assign the values 0 and 1 for *each category* so that 1 indicates the presence of the category and 0 represents its absence. When categorical variables have more than two categories, then more than one dummy variable must be created. However, when creating dummy variables, *never* create more than the total number of categories *minus one*.

• A **dichotomy** is anything that consists of only two values.

Basic Research Designs

Hypotheses not only tell us where to look and what to expect to see, within moral and practical limits they tell us *how* to look. "How to look" refers to the way the observational step in the research process is carried out—how the research is designed. This chapter examines each of the primary research designs used by social researchers, contrasting the kinds of hypotheses each is best suited to test. Subsequent chapters are devoted to examining each of these research designs in detail.

THE EXPERIMENT

Suppose you want to test this hypothesis: *People will be less likely to vote for female than for male candidates for political office.*

To test this hypothesis, you might examine records of many elections to see whether women are defeated more often than men when they run against one another. The trouble is that, in any such elections, the two candidates will differ in many ways besides their gender—in party affiliation, religion, experience, background, political philosophy, campaign funding, name recognition, and past political record, to name only a few. When women lose to men, is it because people prefer to vote for men or is it a matter of preferring a particular man to a particular woman for particular reasons? Studies done in Canada and in Australia showed that women were more often nominated by their party to run against "unbeatable" incumbents (MacKerras, 1977; Hunter and Denton, 1984).

All of these problems are easily solved by using an experimental research design. An experiment will present people with a choice between candidates who are *identical* except for their gender.

Experiments have two fundamental features: (1) the researchers are able to *manipulate* the independent variable (making it vary as much as they wish, whenever they wish), and (2) there is **random assignment** of persons to groups exposed to different levels of the independent variable. People who take part in an experiment often are referred to as the **subjects** because they are subjected to different values of the independent variable.

MANIPULATING THE INDEPENDENT VARIABLE

To test the preceding hypothesis on gender and voting with an experiment, we could prepare two campaign pamphlets, one for a liberal Democrat and one for a conservative Republican. Two versions of each pamphlet would be created. One of them would be for a candidate named John Green and the other for a candidate named Jane Green. Thus, there would be two pamphlets for liberal Democrats, identical in all respects except for the first name and, therefore, the gender of the two candidates. There also would be pamphlets for conservative candidates differing only in name—Robert White or Roberta White.

Next we would recruit people to take part in the experiments, to be the "experimental subjects." Each subject would be given two pamphlets, one for a conservative Republican and one for a liberal Democrat, and asked to study each and indicate his or her preference. *Which* of the two combinations of party and gender any subject saw would be decided *randomly.* Half of the subjects would be presented with a choice between a conservative woman and a liberal man, and the other half would be shown the reverse combination.

Finally, we would compare the support for each candidate. If there were a shift in support away from female candidates, it would be easy to see and could not be blamed on any other factors since everything else was held constant. If there were no bias against female candidates, the proportion of votes received by the liberal or the conservative candidate should be the same whether the liberal were male and the female conservative or the reverse. If the liberal received more votes when the candidate was male and fewer when female, bias would be the only possible explanation.

When an experiment very similar to this design was conducted by **Laurie E. Ekstrand** and **William A. Eckert** (1981), they found no evidence of gender bias. Although the subjects clearly were inclined to favor liberal candidates, they displayed no significant tendency to prefer male candidates. The results are depicted in Figure 6.1.

Figure 6.1   An Experiment

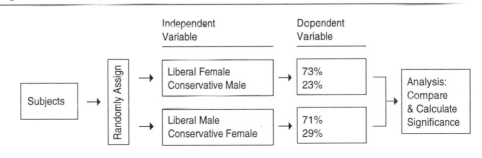

Adapted from Ekstrand and Eckert (1981).

The experiment is the most powerful research design because it so elegantly meets all three criteria of causation. *Time order* is certain—people knew the gender of the candidates *before* they selected one to support. *Correlation* is easily established: Either support shifted as gender shifted, or it did not. But the real power of the experimental method comes from its capacity to rule out sources of *spuriousness.*

RANDOMIZATION

The key to preventing spurious experimental findings lies in the random assignment of subjects. It is decided at random who will be exposed to which value of the independent variable—sometimes experimenters just flip a coin to assign subjects to a group. In medical research, that often means a random method will determine who will get a new vaccine and who will be given a harmless solution instead of the vaccine. In our example, a random method decided who saw which combination of politics and gender.

As is true of any set of human beings, the subjects taking part in experiments differ in many ways. The purpose of randomization is to make the *groups* seeing each combination *alike* so that they will include the same mix of individual characteristics. Thus, each group should include the same percentage of women, of tall people, of well-educated people, and so on. When groups are assigned randomly, it is possible to compute the probability that they are alike. These computations are called *tests of significance*.

SIGNIFICANCE

As was discussed in Chapter 4, a test of significance specifies the odds that a given correlation (or a difference between groups in an experiment) was caused by random variations. Significance tells us when to take an observed result as real and when to reject it as a fluke. Two factors enter into calculations of significance. First is the *number* of subjects in each group—the larger the groups, the less likely that results are random flukes. Second is the size of the observed difference or correlation—the larger the difference or the correlation, the less likely it is to have been caused by a fluke.

Usually, social scientists require that the odds be at least 20 to 1 against a random fluke before they take a result seriously, and they often want the odds to be at least 100 to 1.

In the results of the experiment on gender and voting, the differences were tiny (with there being very slightly more support for the liberal female than the liberal male) and the groups were large—about 250 people in each. We can be sure that these results are real and that it is extremely unlikely that the groups differed other than in the value of the independent variable to which they were exposed.

To sum up: The essential element of all experiments is the same. The experimenters can *manipulate* the independent variable; that is, they can make the independent variable take whatever value they wish, while holding everything else constant. *Randomization* holds everything else constant (the age, race, education, and other aspects of individuals) by making the composition of the groups the same. For this reason, social scientists ought to use the experimental design whenever possible. Chapter 10 will assess basic experimental methods.

Unfortunately, it is impossible to use experiments to study much that social scientists must, or wish to, study. It is impossible to manipulate such important independent variables as sex, race, age, and religion. We can't randomly assign people to be male, white, 37, and Catholic or to be female, Asian, 62, and Baptist. We must take these variables as they occur in the world. Moreover, there are many factors we *could* manipulate but must not. We must not test theories of child-rearing practices by randomly deciding that some children will be neglected or abused and then comparing them with children who were not. So social scientists are forced to use a number of nonexperimental research designs.

SURVEY RESEARCH

In November 1935, in the midst of the Great Depression, the U.S. Congress began debating the annual defense appropriations bill. Seeking funds to spend on relief programs, Democratic Party leaders began demanding very substantial reductions. And, despite the huge military buildups underway in Nazi Germany as well as in Italy and Japan, the news media overwhelmingly supported defense cuts, too. Moreover, members of Congress and the news media presented the cuts in the form of a national mandate, claiming that the "American people" *demanded* that the armed forces be reduced to a "skeleton force" and the money be used to help the unemployed. Even those who opposed military cuts believed that most people wanted them.

In previous years, these perceptions of public attitudes would have gone unquestioned. But something important had changed: During the week of September 10, 1935, the Gallup Poll had conducted its first-ever national opinion survey. Having successfully sold newspapers throughout the country the right to publish their results, the Gallup Poll continued sending its interviewers out every week to interview a national sample of about 1,400 adults (a new group each time) about current issues. The week of December 1, 1935, the Gallup interviewers asked, "Do you think that army appropriations should be greater, smaller, or about the same as they are now?" The question was repeated to ask about navy and about air force appropriations. The results were these:

	Army	Navy	Air Force
Greater	48%	54%	74%
Smaller	11%	11%	7%
About the same	41%	35%	19%
	100%	100%	100%

Hardly anyone wanted cuts in defense spending. Reading across the second line in the table we see that only 11 percent wanted to cut spending for the army and the navy, and only a minuscule 7 percent wanted cuts in air force funding. Reading across the top row, we see that 48 percent wanted to increase army funding, 54 percent wanted more for the navy, and 74 percent said that more should be given to the air force. So much for expert views about public opinion.

If you wish to know the distribution of some trait in a population, whether it is a large nation or a small village, it is necessary to collect information either from everyone or from an adequate sample of that population. Questions about how many people think one way or the other or about what factors cause people to differ in their opinions are simply beyond the capacity of "experts" to gauge accurately. To continue our example, because the only substantial opposition in Congress to defense cuts came from a few Republicans, it was assumed that this also was somewhat of a partisan issue in the public at large. Gallup reported otherwise, as shown.

Air force appropriations should be:

	Democrats	Republicans
Greater	75%	72%
Smaller	5%	9%
About the same	20%	19%
	100%	100%

Reading across the top row in the table we see that 75 percent of Democrats and 72 percent of Republicans favored more money for the air force. These differences are trivial and fall far short of statistical significance.

Survey research is based on **samples** of individuals who are **interviewed** or who fill out **questionnaires**. People included in the survey often are referred to as **respondents** since they respond to the questions asked by the interviewer or included in the questionnaire.

STUDYING "NATURAL" VARIATION

Surveys permit us to study many variables that are impossible or immoral to manipulate. That is, surveys permit us to examine variation as it naturally occurs. While we cannot manipulate political party preferences, we can determine each person's preference and then compare the Democrats with the Republicans as the Gallup Poll researchers did. We cannot randomly assign people to live in various regions of the United States, but information on where each respondent lives usually is included in the survey—the analytic technique is to compare people living in different regions. Here's what the Gallup interviewers discovered about regional differences in support for naval appropriations in 1935:

Navy appropriations should be:

	East	South	Midwest	Mountain	Pacific
Greater	62%	57%	40%	43%	59%
Smaller	9%	19%	16%	10%	9%
About the same	29%	33%	44%	47%	32%
	100%	100%	100%	100%	100%

Region mattered a lot. Notice that, in regions with sea coasts (East, South, and Pacific), respondents were substantially more likely to want greater spending on the navy than were those living in the land-locked Midwest and Mountain regions. But, even in the Midwest and Mountain regions, only tiny minorities wanted actual cuts in the navy's budget.

Surveys not only can study independent variables as they naturally occur in the world and independent variables that cannot be manipulated, but they also are used to study natural dependent variables. For example, there is an immense literature in political science and sociology devoted to voting research, to testing hypotheses about why different kinds of people vote for or against particular candidates or parties. The Gallup Poll examples we just examined were among the first instances of this research tradition—a tradition that soon spread throughout the democratic nations.

SHORTCOMINGS OF SURVEY RESEARCH

If the great strength of survey research is the capacity to study the world as it is, its primary weakness has to do with demonstrating causation—a weakness found in all nonexperimental research. It is not hard to meet the test of correlation—you already have learned how to do that in the exercises. Often, there is some ambiguity about time order, but a careful researcher usually can sort things out. However, the possibility of spuriousness always persists. You have learned several methods by which social researchers attempt to eliminate spurious correlations. But these methods never provide definitive proof that a given correlation is not spurious—because we never can exhaust all possible sources of spuriousness.

Perhaps the concern most commonly expressed about survey results has to do with the accuracy of the data gained by asking people questions. The issues are these: *Will* people give honest answers and *can* they do so? An immense amount of research has explored the honesty of responses, and many techniques have been developed to increase honesty. Clearly, there are pitfalls to avoid in designing questions for use in surveys. Chapter 7 will deal in some detail with these as well as other basic steps in doing survey research.

FIELD RESEARCH

While attending graduate school in the sociology department at the University of California, Berkeley, **John Lofland** and **Rodney Stark** (1965) became interested in understanding the process by which people convert to new religions. Having reviewed the literature, they discovered that virtually nothing had been written about this topic and what little there was seemed suspect—the writers seemed to lack any firsthand knowledge. So Lofland and Stark could formulate a research question: Why and how do people convert? But they couldn't state a hypothesis because they didn't know enough about conversion. However, they did know enough to recognize the proper research method to use.

Field research involves going out to observe people as they engage in the activities the social scientist wants to understand. This method is called *field research* because the research is conducted in the field—in the natural settings in which the people and activities of interest are normally to be found. Sometimes, field research is guided by hypotheses, but often it is exploratory and produces suggestive hypotheses only after the observations are completed.

In pursuit of their research question, Lofland and Stark searched for a new religious group that was gaining converts. Eventually, they found a group of 12 people who had just moved to San Francisco from Eugene, Oregon. The group was led by a missionary from Korea, Dr. Young Oon Kim, who had been sent to convert Americans and Canadians (she had an advanced degree from the University of

Toronto) to a new religion founded by a Korean electrical engineer, Rev. Sun Myung Moon. What the sociologists had stumbled upon was the first group of American members of the Unification Church—widely known today as the Moonies.

Lofland and Stark began attending the group's religious services, which were held in the evening in the apartment building occupied by the group. After a few weeks, the sociologists were sure that this group was exactly what they needed and so they settled in to watch as new people came in contact with the group. What they hoped to do was identify factors that determined who among these people did or did not convert. At this point, Lofland and Stark shifted from *covert* to *overt* observation.

COVERT OBSERVATION

Their initial contact with the group was the result of accepting an invitation to come and hear the new religious message. While they made no secret about being sociologists, Lofland and Stark said nothing about studying the group. In that sense, they were covert observers who waited until they were away from the group before recording their observations.

Field researchers engage in **covert observation** when those they observe are unaware that they are the objects of research.

The advantage of covert observation is that it minimizes observer effects. That is, the Moonies treated Lofland and Stark the same way they treated other people whom they wanted to convert because that's how they viewed them. Hence, the presence of the sociologists did not disturb normal patterns of behavior. The disadvantage of covert observation is that the observers must act as normal participants and this can interfere with the research. For example, they are very limited in how inquisitive they can be. In this instance, as potential converts Lofland and Stark couldn't conduct interviews or openly take notes about what was happening when they visited the Moonies.

OVERT OBSERVATION

Lofland and Stark eventually sat down with Dr. Kim and explained that they wanted to do a study of the group in order to understand the process of conversion. They were careful to be sure she understood what that would involve: They would want to interview people at length and to attend meetings, and eventually their results would be published. Dr. Kim told them to go ahead. If nothing else, she explained, the study might serve as a detailed historical record of the early years of the movement in America.

When those being observed know they are objects of field research, this is known as **overt observation**.

Because members already knew Lofland and Stark quite well, the shift to overt observation went smoothly. Moreover, since Lofland and Stark had observed a lot of normal behavior as covert observers, they were able to be confident that their presence as known observers wasn't causing people to act differently.

FIELD NOTES

Experiments produce written records of how each subject in each condition responds. Surveys result in huge computer data files in which each respondent's answers to all questions are recorded. Field researchers write notes about what they see and hear. These constitute the data from which conclusions are drawn. In Chapter 9, we will see some examples of field notes and find out how some field researchers recently have begun to computerize their notes and how these qualitative data sometimes can be quantified.

As they studied their field notes, Lofland and Stark reached the conclusion that conversion is mainly the result of interpersonal attachments. No one joined the Moonies as the result of a religious search—most converts had not even been all that interested in religion before their conversions. Instead, *people joined if and when their attachments to members of the group outweighed their attachments to those outside the group.* Many converts were old friends or relatives of people already in the group. The others were people who formed close friendships (and sometimes romantic ties) to group members. In the end, conversion was an act of aligning one's religious behavior with that of one's family and friends.

Since they published their findings, this conclusion by Lofland and Stark has served as a hypothesis in many other field studies of religious groups and even has been tested with survey data. By now, more than 25 such replication studies have been conducted, and all of them have confirmed the hypothesis (Kox, Meeus, and t'Hart, 1991).

Field research has several limitations. First, it is best suited for the study of relatively small groups. Lofland and Stark could keep a close eye on the first 12 Moonies and on converts as they came in one-by-one. But, as the Moonies grew, eventually it became impossible for a field observer to be sure of the distribution of various attitudes, beliefs, and background traits within the group. So Eileen Barker conducted a survey to supplement her observations—as we shall see in Chapter 9, which also supplies details on how to do field research. The second limitation is common to all nonexperimental designs. Lofland and Stark couldn't offer proof that the correlation between attachments to the group and conversion was not a spurious relationship.

AGGREGATE OR COMPARATIVE RESEARCH

For many important social research questions, individuals are the wrong unit of analysis. You can't explore why democracy thrives in some nations and not in others by doing studies of individual political attitudes, or understand why murder rates are high in the United States and low in Canada simply by interviewing Americans and Canadians. Questions about aggregate units can be answered only by examination of such units. Research based on such units often is called *comparative* research.

All research is based on comparisons. To see the results of an experiment, we compare the results for groups subjected to different values of the independent variable. In surveys, we also compare across categories or levels of the independent variable. However, the term **comparative research** usually identifies comparisons of large social units, typically whole nations or societies, but sometimes states, cities, or counties (Ragin, 1987).

This use of the term *comparative research* began in the nineteenth century, long before social scientists did surveys or even many experiments, back when some social scientists attempted to compare several societies in an effort to account for differences among them. These studies typically compared no more than three or four European nations, and the research topics ranged from why some nations had higher suicide rates than did others to why some had proportionately more beggars. While these studies often presented much quantitative data, there was little attempt to analyze them since few statistical techniques were yet known.

Today, the journals are filled with quantitative and statistically sophisticated studies based on scores of nations and societies.

DATA "COLLECTION"

The primary difference between what comparative researchers do and what other social researchers do involves how they obtain their data. In an important sense, the first three research designs we have examined in this chapter are basic methods of data collection. Experimenters record what was done to each subject and how he or she responded. Survey researchers interview samples of respondents or ask them to fill out questionnaires. Field researchers go out and write down what they see people doing. But comparative researchers spend most of their time in the library or at their desks because they depend primarily on other people, especially government agencies, to collect their data for them.

USING OFFICIAL DATA TO COMPARE NATIONS

Look in any almanac and you will find about half a page devoted to each nation in the world (except for a few tiny ones). Each of these sections is crammed with statistics. Most of these statistics come from government reports, although some of the data are collected by private organizations that then report them to the government. The latest edition of *The World Almanac and Book of Facts,* for sale in any supermarket, includes such statistics for each nation as population size and density (the number of inhabitants per square mile); ethnic, religious, and language composition; area; life expectancy; infant mortality rate; percent employed in agriculture; rates of TV, radio, and telephone ownership; GNP; number of physicians and hospital beds; the literacy rate; and much more. *The World Fact Book,* published annually by the U.S. Central Intelligence Agency (on sale in most larger bookstores), includes an even more complete set of statistics including the number of miles of paved highways and the size of each nation's military forces. Each year, INTERPOL (the International Police Organization, headquartered in France) issues *International Crime Statistics* covering most nations. The World Bank supplies annual reports on economic development and on each nation's international trade. The United Nations also issues annual statistics including the *Demographic Yearbook,* which offers a variety of health statistics including causes of death as well as data on many cultural activities such as book publishing. The Population Council issues annual reports on fertility and on the use of contraceptives, while groups such as Freedom House report yearly on issues of political freedom and human rights. This list includes only a few of the best sources of data.

Clearly, a great deal of comparative research can be, and is, based on these data, with nations as the units of analysis. Let's briefly examine such a study.

Glenn Firebaugh and **Frank D. Beck** (1994) wished to test the dependency theory of economic development, which claims that foreign investment in Third World, or less developed, nations makes those nations dependent on developed nations. The advanced nations exploit this dependency through multinational companies with the consequence that economic growth is impeded in the less developed nations and the development that does occur benefits only the rich.

To test this theory, Firebaugh and Beck assembled data on economic growth, trade, foreign investment, quality of life, and economic inequality from published sources for 62 nations, covering the period from 1965 through 1988. Their results failed to confirm the predictions from dependency theory. Foreign investment and trade typically increase economic growth in less developed nations, and such growth results in an improved quality of life for the poor masses, not only for the wealthy.

CROSS-CULTURAL "SAMPLES"

Anthropologists pioneered field research methods. For nearly two centuries, they have been conducting ethnographic (or descriptive) field studies of human societies (Evans-Pritchard, 1981). Usually, the emphasis has been on recording detailed accounts of the culture of preliterate societies—an effort meant to preserve this information before the society was destroyed or its original culture was changed beyond recognition through contact with more developed societies.

Besides preserving knowledge of cultural patterns, these ethnographies have served as the basis for comparative studies that seek to demonstrate how various aspects of culture "fit together"—for example, to show that, in societies where women are more important in providing subsistence, there is less gender inequality.

For a long time, the method most often used in such comparative studies was to compare one or several societies having a high level of female involvement in obtaining subsistence with one or several societies where women contribute little to subsistence and then to show that they differ in terms of gender equality, for example.

However, as large numbers of ethnographic accounts of different societies began to pile up, some researchers began to propose that such comparative studies be based on many more cases and that statistical methods be used to analyze the data (Kluckhohn, 1939). To do so, however, required that the field reports written by many different ethnographers about many different societies be systematically transformed or coded to create a uniform set of variables in numerical form. For example, lengthy accounts of subsistence activities had to be converted into a set of scores that would constitute ordinal variables, including such variables as how dependent each culture is on hunting and gathering for subsistence.

After World War II, a group of anthropologists at Yale, led by **George P. Murdock**, began a vast project to code a data set based on ethnographic accounts in which simple societies were the units of analysis. The initial work was done under the auspices of the Human Relations Area Files, Inc. Subsequently, Murdock moved to the University of Pittsburgh where the coding operation continued and a journal was founded to publish the results and to promote use of the data—*Ethnology*. In its first issue, *Ethnology* began to publish installments of coding based on field studies of hundreds of societies. By 1971, data on 1,264 cultures (societies) had been published, a data set referred to as the *Ethnographic Atlas*.

By then, however, Murdock and his associates had become concerned that many of the cases included in the data set were poorly documented. Worse yet, it seemed that many of them were "duplicates" in the sense that many cultures appeared several times because ethnographers had studied different villages or subgroups belonging to the same society and having essentially identical cultures; by appearing multiple times in the data set, one case might greatly influence the results. So, in 1969, Murdock and **Douglas R. White** published an initial set of variables for a subset of cases which they identified as the *Standard Cross-Cultural Sample*.

It might have been called the Standard Sample of Preindustrial Societies, but anthropologists typically use the word *culture* to refer to groups other social scientists would call *societies*. Perhaps the set shouldn't have been called a "sample," since, technically, these cases were not a sample of a well-defined universe. However, because Murdock invested so much effort in selecting a "representative" set of societies from the ethnographic literature, social scientists treat the data as if they were a true sample. In any event, the *Sample* was a substantial improvement over the *Ethnographic Atlas*. It includes 186 cases worldwide and was created to avoid duplications, to give substantial representation to each continent, to drop poorly documented cases, and to limit the cases to preindustrial societies. Although the cultures included come from different periods in human history (the ancient Hebrews are one of the cases, for example), data on each case are from the same point in time. That is, all codes for the ancient Hebrews date from Old Testament times, the data for the Aleut of North America date from 1840, and those for the Pawnee Indians apply to the late 1860s. Figure 6.2 offers examples of these cross-cultural data.

Murdock still was not satisfied. So, in 1981, he decided to carve a new "sample" from the *Ethnographic Atlas*. His purpose was to have enough cases to permit analysis within the major geographic regions of the world. This time, he selected 563 societies to make up what he called the *Atlas of World Cultures*. Here, too, Murdock took pains to eliminate duplicates, avoid poorly documented cases, and limit the cases to preindustrial societies. The regional distribution of these societies is as follows:

Sub-Sahara Africa	111 societies
North America	124 societies
South America	81 societies
East Asia	81 societies
Insular Pacific	101 societies
Circum-Mediterranean	65 societies

Figure 6.2   A Cross-Cultural Data Sampler

From the *Standard Cross-Cultural Sample* (186 Societies)

Percentage of subsistence labor supplied by females

	Frequency	Percent
Under 10%	15	8
10%–25%	42	23
26%–50%	104	57
Over 50%	22	12

No data for 3 cases

Do they have a written language?

	Frequency	Percent
No writing	73	39
Pictorial	69	37
Writing	44	24

From the *Atlas of World Cultures* (563 Societies)

Degree of dependence on hunting for subsistence

	Frequency	Percent
None	171	30
Some	179	32
Much	213	38

Importance of slavery in this society

	Frequency	Percent
Absent	307	57
Some	117	22
Much	111	21

No data for 28 cases

From the North American subset of the *Atlas of World Cultures* (124 Societies)

Degree of dependence on hunting for subsistence

	Frequency	Percent
None	2	2
Some	15	12
Much	107	86

Importance of slavery in this society

	Frequency	Percent
Absent	82	68
Some	25	21
Much*	14	11

No data for 3 cases

*Most people are unaware that slavery was common among Northwest tribes, who enslaved members of other tribes captured during raids.

Since Murdock created these databases, other scholars have been adding variables to these data sets—especially to the Standard "Sample." For example, in 1983, **Gwen J. Broude** and **Sarah J. Greene** added data on wifebeating to 70 of the societies included in the Standard "Sample." They coded a "significant" amount of wifebeating in 80 percent of these societies, while finding it to be rare in 20 percent. For a few years, however, no attempt was made to use these data to see how the societies with a lot of wifebeating differed from those having only a little.

Gerald M. Erchak and **Richard Rosenfeld** (1994) found that wifebeating is part of a coherent pattern of cultural norms sustaining violence. That is, societies constantly engaged in external warfare, having frequent outbreaks of internal strife and violence, and which define masculinity in terms of a "warrior ethos" are the societies having very high rates of wifebeating.

As noted, researchers have continued to add variables to these cross-cultural data sets. However, as in the case of wifebeating, too often scholars have chosen to save time and effort by adding new codes for only a subset of the cases. In doing so, however, they seem unconcerned that they have, in effect, created a new, smaller "sample." Moreover, there has been no consistency in selection of a subset of cases to code. Thus, one researcher may add variables for each evenly numbered case, while another will add variables only to the odd-numbered cases, with the result that no case is coded by both researchers. That means, of course, that the variables added by the first researcher cannot be analyzed in connection with the variables added by the second researcher. Nevertheless, an immense number of potentially fruitful studies remain to be done with these databases.

CONTENT ANALYSIS

In a sense, Murdock's cross-cultural data sets are based on **content analysis**, the research method whereby verbal and nonquantitative textual and graphic materials are coded to produce quantitative data. That is, the anthropologists who wrote the ethnographic studies Murdock and his colleagues coded were not consciously rating a society on a set of variables. They provided detailed descriptions of what they saw, and Murdock imposed a set of measuring criteria on these reports in order to create variables. However, these materials could not have been coded so easily in this fashion had anthropologists not had a research agenda based on comparisons. That is, it was well established what observations each ethnographer was expected to make—the set of information each was expected to produce through his or her field study.

However, many verbal and textual materials are produced without the producers having any thought about subsequent use or analysis. Nevertheless, some very important research results have been produced by coding such material to create variables.

John C. Merrill (1965) believed that *Time* magazine had been biased in its coverage of three successive American Presidents: Harry S Truman, Dwight D.

Eisenhower, and John F. Kennedy. He randomly selected 10 issues of *Time* published during each president's time in office. Then he had six judges independently examine and code each story in which the president's name appeared. Among the categories were biased adjectives, verbs, and adverbs: "Truman snapped"; Eisenhower's "warm manner of speaking." Bias could, of course, be either positive or negative. The judges even coded the photographs: Was the president depicted as calm/nervous; dignified/undignified; angry/happy?

The judges agreed on 93 instances of bias toward Truman, 92 of them negative. In contrast, of 82 instances of bias toward Eisenhower, 81 were positive. They identified 45 instances of bias toward Kennedy, 31 of them positive.

When the writers and editors at *Time* wrote and edited these articles, they were unaware that one day their work would be transformed into quantitative data, but it seems likely that they knew very well the magazine supported Eisenhower and opposed Truman.

Because Merrill published full details on his methods, it would be possible to repeat and extend his study to include more recent presidents. It might also be interesting to compare several magazines, particularly in their treatments of the same story. And it also would be possible to code the coverage received by presidents throughout their terms to see if the coverage shifted from positive to negative as was probably the case for George Bush and for Bill Clinton.

STUDYING CHANGE

Changes over time are central to social research. But, rather than constituting a basic research design, each of the designs covered in this chapter has special applications devoted to studying change.

Through the years, many experiments have been devoted to studying attitude change or to evaluating the effects of various policies, programs, products, or procedures on the basis of before-and-after comparisons.

Data from surveys taken over a period of years often are used to see trends in opinion or even in activities such as attending church (Greeley, 1989). Sometimes, survey researchers reinterview the same respondents several times in an effort to identify trends—this often involves studies of change in support for candidates over the course of an election campaign. Occasionally, data are collected by reinterviewing the same people over many years.

Field researchers often report changes they observed as they occurred—sometimes a researcher will revisit the same group many times over many years (Chagnon, 1988). Sometimes, a group has been restudied by a different researcher, often many years after the original fieldwork was conducted (Barker, 1984).

But perhaps a greater proportion of comparative research has been devoted to studies of change than is true for any other method. The immense literature on industrial development, for example, is a literature about social change.

In the chapters that follow, these special applications for the study of change will be dealt with in detail.

CONCLUSION

The purpose of this chapter was to offer an overview of the remainder of the book—to let you see the large contours of social research methods before moving closer to examine the details.

Clearly, there are many ways to do social research, and many social scientists soon come to prefer one method over another. While this is entirely understandable, it is somewhat limiting in that the "right" way to do research depends on what you want to find out. If you want to know what happens when potential converts begin to interact with missionaries from a religious group, the "right" method is to go out and see, not to distribute questionnaires to those who already have converted. But, if you want to know whether gender enters into most voters' preferences, looking is no substitute for an experiment. To discover whether underestimates of the probability of being caught encourage delinquency, a survey is the "right" method. And, if you wish to know whether economic development leads to decreased economic inequality, only the comparative method is appropriate. The only qualification to matching methods to hypotheses is that ethical concerns often require that a less appropriate method be used. An experiment probably would be the best way to test theories of conversion to unusual religious groups, but no legitimate social scientist would conduct such an experiment. Hence, ethical issues associated with each method will be discussed in each subsequent chapter.

REVIEW GLOSSARY

- **Experiments** have two fundamental features: (1) the researchers are able to *manipulate* the independent variable (making it vary as much as they wish, whenever they wish), and (2) there is **random assignment** of persons to groups exposed to different levels of the independent variable. People who take part in an experiment often are referred to as the **subjects** because they are subjected to different values of the independent variable.

- **Survey research** is based on **samples** of individuals who are **interviewed** or who fill out **questionnaires**. People included in the survey often are referred to as **respondents** since they respond to the questions asked by the interviewer or included in the questionnaire.

- **Field research** involves going out to observe people as they engage in the activities the social scientist wants to understand. This research is called *field research* because the research is conducted in the field—in the natural settings in which the people and activities of interest are normally to be found. Sometimes, field research is guided by hypotheses, but often it is exploratory and produces suggestive hypotheses only after the observations are completed.

- When those being observed are unaware that they are the objects of research, researchers are engaging in **covert observation**. When the identity of the observers as social researchers is known to those being observed, researchers are using **overt observation**.

- While all research is based on comparisons, the term **comparative research** usually identifies studies based on aggregate units of analysis—nations, states, cities, counties.

- **Content analysis** is the research method that consists of coding verbal and non-quantitative textual or graphic materials to produce quantitative data.

Survey Research

- **STUDYING CHANGE**
  - TREND STUDIES
  - PANEL OR LONGITUDINAL STUDIES
  - AGE, COHORT, AND PERIOD EFFECTS
- **DATA ARCHIVES**
- **ETHICAL CONCERNS**
- **CONCLUSION**
- **REVIEW GLOSSARY**

Although surveys of a small and concentrated population can be done on a modest budget, surveys of large, dispersed populations are expensive. It can cost $2 million or even more to conduct a national survey of the United States. Obviously, social scientists would not conduct surveys if there were a cheaper and equally accurate way to obtain the same information. But, for particular sorts of questions, the only comparable method is a census, which is even more expensive to conduct.

To more fully grasp the need for surveys, it will be helpful to recognize the distinction between simple and proportional facts.

SIMPLE AND PROPORTIONAL FACTS

A **simple fact** is an assertion about a concrete and limited state of affairs, often merely the claim that something happened or exists and usually having to do with only one or very few cases. Some examples will help:

- This tribe holds a feast every full moon.

- Some Canadian mothers nurse their babies.

- Jack Wong is a Chinese American who lives in Chicago.

A **proportional fact** asserts the distribution of something, or even the joint distribution of several things, among a number of cases. Questions requiring proportional facts include:

- How many tribes hold full-moon feasts?

- What proportion of Canadian mothers nurse their babies and, on average, for how long do they do so?

- What percentage of people in Chicago are Chinese Americans and how many have non-Chinese first names?

It often is relatively cheap and easy to establish simple facts. Any reliable informant can provide an anthropologist with information about regularly scheduled feasts in his or her society. Knowing a few Canadian mothers who do nurse their babies is sufficient. And a chance meeting with Jack Wong in a waiting room at O'Hare Airport could establish the third simple fact.

But no experts can provide answers to any of the questions concerning proportional facts *unless* they have conducted a census or a survey.[1] Thus, to the extent that proportional facts are the focus of attention, the costs of survey and census studies must be borne. That does not mean, however, that surveys can yield reliable information on just any subject.

---

[1] To determine how many tribes hold full-moon feasts, it would be necessary to examine data from a sample of tribes (however defined) or from all of them.

RELIABILITY

Recall from Chapter 3 that lack of reliability involves poorly or inaccurately measuring a concept as opposed to a lack of validity, which has to do with failing to measure the correct concept. Three main factors cause unreliability in survey research data. The first is asking people questions they can't answer or can't answer accurately. The second is asking people questions they won't answer or won't answer honestly. The third has to do, not with how individuals answer, but with the proportion of people who refuse to take part, thus causing some unreliability in the overall distributions of responses. A later section of this chapter will be devoted to the problem of unreliability due to nonresponse. Sections on constructing survey questions will deal with the first two problems in detail. However, it will be useful to begin with an overview.

CAN'T ANSWER

Survey respondents often give unreliable answers because they really don't know the answers. For example, research shows that most people don't really know how many hours of television they watch on an average weekday. If you want only a rough estimate so that people who watch a lot can be compared with people who watch little or no TV, then it is sufficient to ask them to estimate their viewing time. If greater precision is needed, it is necessary to get people to keep a log of their viewing for a period of time.

As another example of this problem, many respondents have difficulty with questions about income. Sometimes, this is because the question asks about their annual income and they know only their hourly, weekly, or monthly income. Some people report their gross income, while others report their net or take-home income. Questions about annual family income present even greater problems. Some respondents don't know how much other family members earn.

Respondents are especially prone to unreliable responses when the question is essentially irrelevant to them. In the last few months before presidential elections, the Survey Research Center at the University of Michigan conducts a series of national election studies. Sometimes, these national surveys reinterview the same respondents several times in an effort to chart shifts in voter attitudes and choices during the course of the campaign (see the discussion of panel studies later in the chapter). In a classic study analyzing these data, **Philip E. Converse** (1964) found that about half of the population really didn't have opinions on many of the issues they were asked about. That is, the best description of their response patterns from one interview to the next was *random*. They would say they were in favor of some political proposals in the first interview and say they opposed them in the next interview in an entirely helter-skelter way. They did this, not because they were stupid, but because they felt obliged to answer questions about issues that they didn't care about one way or the other.

Finally, unreliable answers often result from misunderstanding the questions. The principal reason is that some questions are so badly formulated that it is difficult or even impossible to know how to answer them accurately. Many survey questions—even in surveys conducted by some of the most reputable organizations—defy an answer. Even worse, they can elicit answers that completely mislead. The immense anguish caused by an incompetent question about the Holocaust is an edifying case.

In 1993, the Roper Poll reported that 22 percent of American adults doubted that the Nazi extermination of Jews during World War II ever happened. Worse yet, another 12 percent said they didn't know if it had happened or not. Understandably, this report stunned the Jewish community, and the press quoted many prominent Jewish leaders about the need to somehow educate the public. Here is the question asked by the Roper interviewers:

> Does it seem possible or does it seem impossible to you that the Nazi extermination of the Jews never happened?

Given the double negative, it is hard to know what this sentence means or how to interpret an answer. It is even harder to imagine how an experienced and reputable firm such as Roper could have asked such a question. But it soon became clear that most people who seemed to agree that it seemed possible that the extermination never happened actually had no doubts that the Holocaust happened but were merely confused by the strange wording. We know this because the Gallup Poll asked a very clearly worded question and found that fewer than 0.5 percent of Americans said the Holocaust "definitely" did not happen, while another 2 percent said it "probably" did not happen.

In the wake of these results, Burns Roper apologized to his colleagues in the American Association of Public Opinion Research, "This is not the note on which I wanted to conclude my 48-year career in the opinion research field." He went on to regret that this badly written question "served to misinform the public, to scare the Jewish community, and to give aid and comfort to the neo-Nazis who have a commitment to Holocaust denial" (Kifner, 1994). But more than a year passed between the publication of these sensational results and their rebuttal.

WON'T ANSWER

A second major source of unreliability in survey research is asking people questions they won't answer or won't answer honestly. For example, when early surveys produced age distributions with too many people in the age groups 29, 39, 49, 59, and 69 and too few people in the 30, 40, 50, 60, and 70 age groups, it was recognized that some people were unwilling to admit they had passed one of the milestone ages and delayed doing so. To overcome this problem, survey researchers began to ask people their year of birth, rather than their age. The problem disappeared.[2]

---

[2] In recent years, this tendency has become sufficiently rare so that once again surveys simply ask respondents their age.

We have noted that often people lack adequate information to report their income accurately, but questions about income also suffer from the unwillingness of some respondents to tell what they earn and the tendency for some respondents to overreport their income—often in order to impress an interviewer. The same two factors probably influence the reliability of responses to questions about sexual activities: Some people won't tell and some people brag.

A solution to asking very sensitive questions involves having the interviewer hand the respondent a questionnaire and an envelope. The respondent completes the questionnaire, seals it, and hands it back to the interviewer with the assurance that it will be opened only by data entry clerks who will have no idea whose questionnaires they are reading.

VALIDITY: ATTITUDES AND BEHAVIOR

The validity of survey research rests on what people say—what they tell us about what they think, how they feel, and how they act. Since validity refers to whether or not we are measuring the concept we intend to measure, we must be careful not to confuse measures of what people say they think and feel with measures of what they say they do. Put another way, we must know when we are measuring *attitudes* (or attempting to do so) and when we are measuring *behavior*.

An **attitude** is an enduring, learned predisposition to respond in a consistent way to a particular stimulus or set of stimuli. We all have attitudes toward significant things in our environment: objects, people, ideas, events, and so on.

Attitude is the central idea of social psychology and has generated an immense literature. It usually is regarded as consisting of three components: (1) a belief, (2) a favorable or unfavorable evaluation, and (3) a behavioral disposition (DeLamater, 1992). Notice that the third component is not behavior as such, but a disposition or propensity to act in certain ways on the basis of the attitude. Consider these statements: "I believe that's a pizza. I really like pizza. I will probably eat that pizza." But no eating has yet occurred—eating is not an attitude, it is a behavior.

Behavior involves action; it is what living creatures *do*. We "hold" attitudes, but we "perform" behavior.

A huge number of social scientific *concepts* identify specific attitudes or sets of attitudes: religious belief, trust in government, alienation, racism, support for civil liberties, political conservatism, trust in people, and hundreds more. Consider the concept of *gender bias*—defined as believing that women tend to be less suited than men for various responsibilities. The following item from the General Social Survey is designed to be an indicator of that concept:

Most men are better suited emotionally for politics than are most women.

Agree	21%
Disagree	79%

Those who agreed with this question clearly hold a biased attitude. But the item makes no mention of behavior; that these people might be inclined to vote against female candidates for office is merely implicit—a disposition to act in a particular way. The next question from the General Social Survey attempts to measure gender bias as a behavior, rather than an attitude.

If your party nominated a woman for president, would you vote for her if she were qualified for the job?

Yes	91%
No	9%

It is important to know that there is no one-to-one match-up between the attitude measured by the first item and the anticipated behavior measured by the second. As shown below, the majority of people who thought men better suited for politics nevertheless said they would vote for a qualified woman for president:

Men better suited:

Vote for a woman?	Agree	Disagree
Yes	69%	97%
No	31%	3%

An immense amount of research and theorizing has been devoted to the link between attitudes and behavior, since substantial slippage between the two, like that above, frequently is observed (Fishbein and Ajzen, 1975; Kiecolt, 1988; DeLamater, 1992). There is consensus that attitudes do predict behavior, if imperfectly. However, it would be inappropriate in a methods textbook to pursue so complicated a substantive matter at greater length. It is sufficient to remember two points:

1. Attitudes are not measures of behavior.

2. Questions that ask about how a respondent *might* act in a hypothetical situation are only *weak* measures of behavior.

None of these respondents ever has faced the option of voting for a woman for president of the United States. Thus, the item above is partly an attitude in that respondents are anticipating what they *would* do, not reporting what they actually have done.[3] Not surprisingly, people are far better able to report how they have acted than to anticipate what they might do.

Let's see how these general concerns about the reliability and validity of survey research data are influenced by how the information is gathered.

---

[3] The most recent research, funded by the National Women's Political Caucus and based on an analysis of 50,563 candidates running for state and national offices since 1972, found no gender bias in American election results (Newman, 1994).

INTERVIEWS OR QUESTIONNAIRES?

Surveys are based on interviews or on questionnaires. Questionnaires are far cheaper. It often is possible to conduct a survey based on questionnaires for several dollars per case—even when it is a national sample. All that's usually involved is printing and postage. Surveys based on face-to-face interviews can cost $400 or more per case. Interviews conducted by telephone are far cheaper than face-to-face interviews but still are far more expensive than questionnaire studies.

Given that face-to-face interviewing is so expensive, one would suppose that it must be a far better technique; otherwise, everyone would use questionnaires. As it turns out, however, while face-to-face interviews do have many advantages over questionnaires, they also have drawbacks that questionnaires do not. Thus, the decision about which method to use does not depend entirely on budget, but involves the subject to be studied and the population to be sampled.

FACE-TO-FACE INTERVIEWS

The first advantage of sending interviewers out to conduct a survey is that interviewers can *help* respondents understand questions and thus can gain data from people lacking sufficient literacy to complete questionnaires.

The second is that interviewers can *probe*. They can ask follow-up questions to clarify responses. For example, when asked their religious affiliation, many Americans answer, "Christian." But it would be inaccurate to settle for that answer, classifying the person as lacking a more specific religious preference. In most cases, these respondents are active members of specific denominations—their use of "Christian" reflects that this term recently has come into use as a generic term to identify members of evangelical Protestant denominations. If people write "Christian" in answer to a questionnaire, we can't know whether they are evangelical Protestants or people without denominational preference. But, when interviewers get the response "Christian" to the question about religious preference, they are trained to probe, "What specific Christian denomination is that?" Usually, the respondent then names a specific group such as "Church of the Nazarene" or "Assemblies of God." Questionnaires can include written probes, of course, but this is more cumbersome than relying on an interviewer to probe only when appropriate.

The third advantage is that interviewers can *observe*. For example, interviewers for the General Social Surveys always are asked to report the following:

> In general, what was the respondent's attitude toward the interview? Friendly, Cooperative, Impatient, or Hostile?

Seventy of the respondents in the 1993 study were reported to be impatient or hostile.

Interviewers have been asked to make many sorts of observations of the people they interview and of their homes (interviews nearly always take place in the home). Sometimes, they have rated respondents' physical appearance, the cleanliness of their homes, the exterior condition of their house and yard, whether there was art on the walls, and so on.

The first weakness of interviewing is that interviewers may *influence responses*—even well-trained and conscientious interviewers. Some of the impatient and hostile respondents reported by the NORC interviewers may simply not have liked the interviewers—perhaps there was a bad mix between a well-educated, middle-aged woman and an unemployed, young, male dropout. A more frequent problem is that, face-to-face with an interviewer, some people may feel the need to brag or show off and thus distort their answers. Or they may edit their answers to avoid admitting embarrassing facts or expressing unpopular opinions.

The second is that interviewers may *cheat*. Sometimes, they supply answers to questions when respondents refuse to give an answer. Sometimes, they fake entire interviews with respondents who turn them down. Sometimes, they simply mark answers they prefer (especially on social and political issues) in an effort to influence the outcome of the study. Consequently, the best survey organizations frequently check up on their interviewers.

TELEPHONE INTERVIEWS

When interviewing is done by phone, the same advantages and weaknesses apply with the exception that telephone interviewers are unable to observe anything more than tone of voice and degree of cooperativeness. Telephone interviewing also has the advantage of being far cheaper. But it has several additional drawbacks.

First, the interview must be considerably *shorter*. People often will submit to several hours of interviewing while sitting in their homes. But very few people will endure such a lengthy interview by phone.

Second, there will be a much *higher rate* of *nonresponse*. One reason is the inability to get past answering machines. Another reason is that a direct personal appeal from an interviewer is harder to resist than one from a faceless person calling on the phone, probably at dinner time.

Third, it is far harder to *combine brief questionnaires* on sensitive material with a phone interview. Even if you arrange to send questionnaires to people by mail, many won't come back.

QUESTIONNAIRES

Most questionnaire studies are conducted by mail, although sometimes they are distributed by hand—to students in a class, persons attending a meeting, or employees of a firm. The distinguishing feature is a set of written questions to be filled out and returned by each respondent.

The first advantage questionnaires have over interviews is that they are much *cheaper*.

Second, they can be much *longer*. Reading is much more rapid than speaking and, therefore, questionnaires can cover many more questions in the same amount of time.

Third, they can ask *more complex questions*. In an interview, a question offering respondents a number of alternatives may need to be repeated several times because, by the time the interviewer has recited the last option, the respondent no longer can remember the earlier options. With written materials, it is easy and quick to glance back up the list.

Fourth, they can ask far *more sensitive questions*. Many respondents are entirely comfortable about confiding in people they will never meet and who have no personal interest in them. But, sitting face-to-face with an interviewer, they become unwilling to answer.

The first weakness of questionnaires is that they must rely entirely on respondent reports *without any interviewer observations*.

Second, they must *substitute elaborate follow-up questions* for interviewer probes.

Third, they are limited to respondents having a reasonable level of *literacy*.

It was long believed that an additional weakness of questionnaire studies was a substantially *lower response rate*. But, in his study of hundreds of surveys, **John Goyder** (1985) found that the differences between return rates for mail questionnaires and interview studies has declined substantially in recent years and that the real cause of the difference lies not in the method of data collection, but in the amount of follow-up effort. Because interview studies usually have had far larger budgets, they more often have been able to devote substantial effort to overcoming nonresponse.[4] Thus, when an aggressive follow-up campaign can be mounted, questionnaire studies do about as well as interview studies in terms of response rates. This is particularly true when the target population is literate and people find the study to be of interest to them personally.

Many surveys are unable to arouse interest in potential respondents. Although many people will agree to answer questions about their brand preferences, their vacation plans, or even their political opinions, they may not find the questions interesting and have no stake in the overall results. Sometimes, however, surveys are devoted to matters of serious interest and concern to most respondents. For example, if you drew a sample of members of the American Trial Lawyers Association and asked them to provide information and opinions on proposals to limit the amount of damages awarded by the courts, and if questions were well designed, the response rate would probably be very high. Or a study of college professors on the issue of tenure would likely reach respondents highly motivated to participate.

---

[4] That journal editors force authors of questionnaire studies to report their response rates, while not requiring this of those using interview data collected by established survey organizations, encourages the perception that questionnaire studies have relatively low return rates by obscuring the relatively high nonresponse rates of interview studies (see Chapter 4).

We thus arrive at a rule that is helpful in deciding between interviews and questionnaires: *The less salient the subject, the greater the need to use interviewers.* Once an interview has begun, people will tend to put up with boring questions that would cause them to put a questionnaire aside. However, all questions will bore respondents if they are so poorly designed that respondents can't find answer categories that adequately reflect their views. Thus, we come to the topic of adequately measuring our concepts with survey questions.

MEASURING CONCEPTS WITH SURVEY QUESTIONS

Indicators are empirical measures of concepts. It is necessary that such measures be both reliable and valid. Reliable measures may or may not be valid, but unreliable measures cannot be valid. Thus, the primary concern in creating or selecting survey questions to measure the concepts that motivate our research is to maximize reliability. Survey questions can take many forms, and each form can suffer from flaws that decrease their reliability. In this section, we explore basic principles of question construction.

STANDARD VERSUS ORIGINAL ITEMS

Hundreds of surveys are conducted in North America every year, and this has been going on for decades. Many of the same concepts are measured in survey after survey. Consequently, there are many standard questions that are repeated in one study after another. The use of standard items has several benefits.

First, they permit *meaningful comparisons* across studies (this is especially important for studies of change, as we shall see near the end of the chapter).

Second, they are of *high quality*. Items become standard because they are better than the alternatives and often are the result of a lot of experience and prior effort. Indeed, some have been subjected to careful tests of validity and are known to measure important social science concepts. Some of these measures have been copyrighted and can be used only by permission and for a fee—this is particularly true of various attitude scales created by social psychologists. However, unless such a copyright is clearly asserted in the literature on the measure, you need not be concerned. Moreover, it is the norm among most survey researchers not to copyright their questions and to encourage others to adapt them to their needs.

Third, using standard items *saves time and effort*.

If you want to examine a lot of standard items, the best thing to do is to write to authors of studies you find interesting and ask for a copy of the questionnaire or interview schedule they used. Or you may wish to obtain copies of questionnaires and codebooks (after a survey is complete, researchers usually create a "book" that includes the complete questions, answer categories, and univariate distributions) from a major survey organization or archive (see the section on archives near the end of this chapter).

The trouble is that standard items do not exist for many of the primary concepts to be measured in most surveys. It is taken for granted that researchers will create their own. As you read the following discussion of how to create good survey questions, keep in mind that original items always should be *pretested* before being used. That is, professional survey researchers round up some volunteers to be interviewed or to fill out a questionnaire in order to discover shortcomings that they then can fix before launching the actual study.

CLOSED VERSUS OPEN-ENDED QUESTIONS

The reason to do surveys is to learn the distribution of some variables within a population (or the joint distributions of such variables). Survey questions to measure variables come in two basic forms: closed and open-ended.

Closed questions force all respondents to select their responses from a set provided by the researchers. **Open-ended questions** permit respondents to answer as they wish, in their own words.

Surveys nearly always consist primarily of closed questions, but both forms have advantages and weaknesses.

The first advantage of closed questions is that they maximize *comparability*. When asked if they like jazz music, for example, respondents to a closed question must select from preset categories, clearly differing in terms of how favorable they are. It seems reasonable to assume that those who selected "I like it very much" have quite similar evaluations of jazz and are far more favorable than those who selected "I tend to dislike it." But, when respondents are asked this same question and are not forced to choose among clearly distinct categories, it usually is very difficult to rank many respondents. For example, which of these is the most and which the least enthusiastic: "It's okay," "Less than I did," "More than rock-and-roll," "Sometimes," or, "I'm starting to"? When we can't rank respondents comparably on a variable, we can't use the data to resolve questions such as "Are older people more (or less) likely than younger people to like jazz?"

The second advantage is that they *need not be coded* after the fact because the categories of closed questions already are coded.

The third advantage is that most respondents find them more *efficient*. When respondents are given the option to use closed categories or write in their own answers, they nearly always choose a category (unless the categories are incomplete or poorly drafted).

There are two potential disadvantages of closed questions.

First, they may *suppress variation*. That is, opinions or behavior may be far more diverse than the preset categories will register. For example, the question asking respondents if they had participated in any sports activity in the last 12 months (see Chapter 3) allowed respondents to answer only yes or no. Of the 56 percent who answered yes, some probably participated in a sports activity nearly daily, and others may have done so only once.

Second, closed questions may *misclassify* respondents because they are worded poorly or because answer categories are ambiguous.

There are four strengths of open-ended questions.

First, they sometimes are used not to gather data so much as to allow respondents a chance to *express themselves* (and discharge any frustrations they might have built up from answering closed questions).

Second, they are useful in *exploratory studies* or in preparation for writing closed questions in order to get some sense of how people vary on the phenomena of interest.

Third, they have been demonstrated to work better with *sensitive topics*. **Norman Bradburn** and **Seymore Sudman** (1979) compared the results of open-ended and closed questions about sexual activity and about alcohol consumption. The closed questions offered eight categories ranging from "never" to "daily." The open-ended questions, of course, offered no frequency categories. In each comparison, the open-ended question elicited significantly greater frequency of sex and drinking. Apparently, the presence of the low-frequency closed categories sent an implicit message to respondents that made them less willing to admit higher frequencies.

Fourth, they can provide *juicy quotes* for use in the final report.

The weaknesses of open-ended questions are that they are *expensive* and *difficult* to code and they often fail to provide a sufficient basis for *comparisons* across cases. **Howard Schuman** and **Stanley Presser** (1981) conducted an experiment comparing the results of open-ended and closed questions. Half of the respondents (selected at random) were asked, "People look for different things in a job. What would you most prefer in a job?" Many respondents replied, "the pay." It turned out that some who gave this answer meant "high pay," while many others meant "steady pay." But, given their answers, it was impossible to distinguish the two quite different meanings. A closed question that included both options provided a far more accurate basis for comparison: "Thus building distinctions into the answer categories can more accurately tap differences among respondents than letting them answer in their own words" (Converse and Presser, 1986).

Clearly, the strengths of closed questions outweigh those of open-ended items, and the weaknesses of closed questions can mostly be overcome. Hence, in the remainder of the chapter, our focus will be on closed questions.

EVALUATING CLOSED QUESTIONS

There are many principles and rules-of-thumb to guide the construction of good closed questions. Some are quite obvious and others less so. All are devoted to the same end: to collect reliable and valid data having substantial variation across cases.

Unbiased Language.   The way questions are worded can have an extraordinary impact on how they are answered. Consider this example (Goleman, 1993). During the 1992 presidential campaign, a questionnaire published in *TV Guide* asked readers to mail in their responses. Among the questions was this one:

Should laws be passed to eliminate all possibilities of special interests giving huge sums of money to candidates?

Yes	99%
No	1%

Soon after, a national survey asked this question:

Please tell me whether you favor or oppose the proposal: The passage of new laws that would eliminate all possibility of special interests giving large sums of money to candidates.

Favor	70%
Oppose	28%
Don't Know	2%

Then, still another national survey asked,

Should laws be passed to prohibit interest groups from contributing to campaigns, or do groups have a right to contribute to the candidate they support?

Prohibit contributions	40%
Groups have a right . . .	60%

Clearly, it mattered a great deal how the questions were worded. Few Americans could be expected to say they weren't against letting special interests give "huge sums" to candidates. In addition, these results were not based on a sample but a form of SLOPS research (see Chapter 4)—readers of *TV Guide* are a population, not a sample, and presumably people who found the slant of the questionnaire to their liking would have been far more likely to take the effort to respond.

Notice that, when a good sample was used and the wording was a bit less loaded so that only "large sums" were at issue, a lot of people opposed. Finally, when the question included mention of the rights of contributors, majority support evaporated.

You may wonder how professional survey researchers could have formulated so biased a question as the first. Usually when such biases appear in surveys done by professionals, it is because the researchers are being paid by partisan groups who intend to bias the results to serve their purposes. In this instance, the *TV Guide* questionnaire was prepared for H. Ross Perot, who ran an on-again, off-again third party campaign for president. Perot used the responses to this and other similarly formulated questions to "prove" that "the people" were on his side. The second question appeared in a survey conducted by a well-known polling firm, the Gordon Black Corporation, to see how much bias was produced by self-selected respondents—to learn how respondents drawn in a proper sample would divide on this matter. But, even though their purpose was to replicate Perot's survey, these experienced professionals couldn't bring themselves to use the Perot item in its fully biased form. The results for the third question were obtained by Yankelovich and Partners, another well-known polling organization.

The lesson to be learned from this example is not only to avoid writing biased questions if you really want to find out about public opinion, but also to be suspicious of media reports about public opinion unless they include information about the sample and about who paid for the survey and the actual wording of the question or questions that were used.

Not all bias is this obvious or intentional. Sometimes, even what seem to be rather modest changes in wording matter a lot. A study of opinion on ways to reform civil damage awards proposed to place a cap or upper limit of $250,000 on the amount of money that can be awarded for "pain or suffering" or other noneconomic damages (Kosnick, 1989). Half of the respondents (selected randomly) were asked if they found that proposal *acceptable*. Of these, 90 percent replied that they found it to be "very" or "somewhat" acceptable. The other half were asked if they *supported* this proposal. Only 27 percent said they "strongly" or "somewhat" supported it.

Or consider the results of these two very similar questions included in recent General Social Surveys:

Are we spending too much, too little, or about the right amount on welfare?

Too Little	17%
Right Amount	26%
Too Much	57%

Are we spending too much, too little, or about the right amount on assistance to the poor?

Too Little	65%
Right Amount	23%
Too Much	12%

The majority want to cut welfare spending *and* to increase assistance to the poor. This really isn't a contradiction if we realize that it is the welfare system that people object to, not its underlying intentions. But, had we based our assessment of public opinion on either item alone, we probably would have drawn the wrong conclusions. Many political observers could have concluded from the first question alone that people no longer have sympathy for the poor. However, had they seen only the second question, they could have supposed that people do not want to cut welfare spending.

It is precisely because biases can so easily creep in that professional researchers so frequently use several items having minor variations in language to measure a particular opinion or attitude.

Clarity.   Good survey items must be clearly worded. The Holocaust item reported earlier in the chapter is a case in point. It is awkward to have a double negative in a sentence, and it nearly always is a serious mistake to have one in a survey question. Try to write answer categories to this question:

> Do you think that McDonald's should have not been found not liable for damages when a woman spilled a cup of coffee in her lap?

How much better to ask,

> Do you think that McDonald's should not have been found liable for damages when a woman spilled a cup of coffee in her lap?

And it would have been even better to drop the negative entirely and ask,

> Do you think that McDonald's should have been found liable for damages when a woman spilled a cup of coffee in her lap?

A major flaw in the clarity of survey items often occurs because researchers are trying to economize and therefore cram two distinct questions into one. The result is what is known as a *double-barreled* question:

> If we can now trust the Russians, do you think we should focus our attention on saving endangered species?
>
>   __ Yes
>   __ No

Does "no" mean you don't want to focus attention on endangered species or that you don't think we can trust the Russians? Separate questions should have been asked.

Mutually Exclusive and Exhaustive Categories. The point of closed item categories is to classify people in comparable ways. This goal is unmet if the categories overlap or if significant categories are omitted:

> Please indicate the category in which your current occupation should be classified.
>
>   Professional
>   Managerial
>   Technical
>   White Collar
>   Skilled Craft

Where does the Ph.D. in chemistry who is president of a drug company belong? Such a person is professional, managerial, probably technical, and possibly a white-collar worker—these categories tend to overlap. In similar fashion, where do factory workers belong? No categories are provided for them or for service workers or farmers. Consequently, these categories not only overlap, they also are not exhaustive—they do not exhaust the necessary range of categories.

Questions without exhaustive categories abound in survey studies. Several years ago, two well-known political scientists sent a mail questionnaire to a national sample of college professors. Included were many items about government spending for various programs. For each, the categories were "Far too little," "Too little," and "About the right amount." No provision was made for respondents to indicate that they thought the government was spending too much on some things. Respondents often wrote in responses such as "Too much," but there was no way to know how many others settled for "About the right amount." However, the poor response rate to this survey suggests that many faculty dealt with the problem by simply not completing the questionnaire.

Unfortunately, lack of balance caused by the failure to provide an exhaustive set of categories is not uncommon. A national survey sponsored by *Time* and CNN asked who was at fault for "gridlock" in the federal government. Respondents were allowed to blame congressional Republicans, President Clinton, or both—it seems not to have occurred to the pollsters to allow respondents to blame congressional Democrats or to reject the existence of gridlock.

Forcing Variation.   A frequent error in question construction is premature collapsing of categories. Once the data are collected, it is easy to merge several categories into one more general category—for example, to merge age from one-year categories to categories such as 20–30 (you will learn to collapse survey items in a workbook exercise). But it is impossible to increase categories after the fact. Several years ago, a study of college drug use included this question:

In the past 12 months, with how many different people have you had sex?

＿＿＿ 0–1
＿＿＿ 2–3
＿＿＿ 4 or more

Clearly, one of the most important categories is zero, but with this item it is impossible to distinguish those who have not had sex during the year from those who have. It also seems likely that the highest category is too collapsed as well. In fact, it might have been best not to provide answer categories to this question, but simply to ask people to write in the number as usually is done for questions about age or number of children.

Many items have a natural tendency to be overloaded at one point in the distribution. Sometimes, respondents pile up in the highest or lowest category. At other times, everyone selects the middle category. The result of pileups is reduced variation. Sometimes, nothing can be done about this problem, since this is the real state of affairs. For example, if 99 percent of a sample of residents of Seattle claim to own an umbrella, there are no techniques for decreasing that percentage since it reflects reality, but it might be useful to ask Seattleites how *many* umbrellas they own.

Often, what seems to be a necessary lack of variance isn't. For example, a national survey based on 11,995 high school seniors offered this item, with these results:

I have been in serious trouble with the law.

Yes   4%
No   96%

These results probably are about correct. Few high school seniors have been in serious trouble (those teenagers who do get into serious trouble tend to drop out of school before they become seniors). On the other hand, the extreme lack of variation in response to this question is due to an inadequate view of the concept of delinquency. Delinquency is not like umbrella ownership in that either you have one or you don't. Instead, delinquency can best be measured as a matter of degree defined by variations in the *seriousness* of an individual's offenses and in the *frequency* of such offenses. Consequently, it would have been better to use a series of questions differing in the seriousness of the offense and asking the frequency of each. For example, surveys used to study delinquency include this question:

Have you EVER taken something from a store without paying for it?

Yes, very often
Yes, quite often
Yes, a few times
Yes, but only once
No, not ever

Notice that the categories are in decreasing order. This is to minimize the number who don't admit shoplifting by making them pass over the highest- frequency answer categories first. That is, respondents tend to pile up in the denial category, so the question is structured to increase admissions of guilt. To complete the measurement of delinquency, other items might ask if respondents had ever committed a burglary, used drugs, or stolen money from their parents.

Not only do people often pile up at the low end of variables, they often also pile up at the high end. For example, for many years, surveys included a question about belief in God. Thus, in 1944, the Gallup Poll (survey #335) for the first time asked a national sample of American adults,

Do you, personally, believe in God?

Yes	96%
No	1%
No opinion	3%

The results were essentially invariant. Through the years, whenever the question was asked, the results always were about the same—at least 95 percent always said, "Yes." But it was obvious to some that this lack of variation was not entirely real, that considerable variation in certainty of belief and in definitions of God must exist within the "Yes" category. So a new question that attempted to force variation was written (Stark and Glock, 1965). The new wording achieved the desired results and soon became the standard item. Here is the question and the responses given to it by respondents in the 1993 General Social Survey:

Which statement comes closest to expressing what you believe about God?

I don't believe in God	3%
I don't know whether there is a God and I don't believe there is any way to find out	4%
I don't believe in a personal God, but I do believe in a Higher Power of some kind	9%
I find myself believing in God some of the time, but not at others	3%
While I have doubts, I feel I do believe in God	15%
I know God really exists and I have no doubts about it	66%

Here, too, respondents must pass over the less socially "acceptable" answers before reaching the category registering certain belief. Moreover, there is far more variation when the question is asked this way, and the data are of a form suitable for regression analysis. For tabular analysis, however, some categories have too few cases and the item would need to be collapsed into fewer categories. But, even then, the largest category would include only two-thirds of the respondents, not 95 percent or more. In a workbook exercise based on this chapter, you will be asked to collapse several variables.

CAFETERIA AND FORCED OPTION QUESTIONS

Often, a survey will attempt to economize by grouping a set of items on the same topic. The General Social Surveys usually ask,

Now we would like to know something about the groups and organizations to which individuals belong. Here is a list of various organizations. Could you tell me whether or not you are a member of each type?

Fraternal groups
Service clubs
Veteran's groups
Political clubs
Labor unions
Sports groups
Youth groups
School service groups
Hobby or garden clubs
School fraternities or sororities
Nationality groups
Farm organizations

> Literary, art, discussion or study groups
> Professional or academic societies
> Church-affiliated groups
> Any other groups

This item can be structured in two ways. The first is to ask respondents to select all options that apply to them. In an interview, they would be handed a card with the choices and asked to indicate each type to which they belong. On a questionnaire, the respondent would be asked to check each that applies. This format is called a *cafeteria question*.

A **cafeteria question** allows respondents to pick and choose from a large selection of responses, indicating only the responses they select (questions of this type are sometimes are called *multiple response items*).

The second, and far better, format requires each respondent to indicate an appropriate response for each option or alternative. Such questions are known as *forced options*.

A **forced option question** requires a response to every option.

Thus, the GSS interviewers read the name of each kind of group and ask the respondent to say "Yes" or "No" about whether he or she belongs to a group of this type. On a questionnaire, answer categories should be provided next to each choice so that respondents can answer for each.

Cafeteria questions are not a good way to measure things. When respondents are permitted to pick and choose, it is impossible to fully interpret the status of unselected options. In our example, you would know that a respondent did belong to any sort of group he or she checked, but you couldn't be sure they did not belong to the options that were not checked. Maybe the respondent accidentally skipped over them. This problem becomes especially acute when opinion items are presented in a cafeteria format:

> Here are some opinions people have expressed about movie theaters. Please check those with which you agree:
>
> ____ Most screens are too small.
> ____ There is far too much talking and noise.
> ____ Special effects are much better in a theater than on videotape.
> ____ The prices are too high.
> ____ The sound is much better than on TV.
> ____ The seats are too small.
> ____ The snack bar is overpriced.
> ____ I like theaters because it is more fun to see a movie with other people.
> ____ I try to avoid new theater buildings.

Did respondents skip an item because they didn't agree with it, or only because they agreed with it less strongly than the ones they marked? That is, does an unmarked item mean *agree a little bit*, *no opinion*, or *disagree*? This ambiguity does not arise if you force respondents to agree or disagree with each statement.

CONTINGENCY QUESTIONS

Often it is appropriate to ask certain questions only if the respondent has answered a previous question in a particular way. For example, if people have just indicated they have never been married, it makes no sense to ask them how old they were when they got married, how happy their marriage is, or how much education their spouse has. Consequently, surveys include many contingency questions. **Contingency questions** divert respondents to different questions or direct them to skip questions, *contingent* on responses to a prior question.

It is very important to structure contingency questions so that interviewers will be accurately led through the alternatives or, with questionnaires, so respondents can follow the proper path. The questionnaire included in Appendix B of the workbook contains this contingency question:

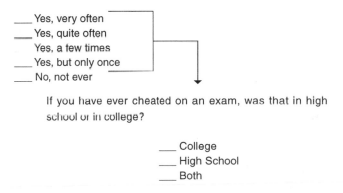

Have you ever cheated on an exam?

____ Yes, very often
____ Yes, quite often
____ Yes, a few times
____ Yes, but only once
____ No, not ever

If you have ever cheated on an exam, was that in high school or in college?

____ College
____ High School
____ Both

ITEM ORDER

It is important to maximize respondent interest in an interview or a questionnaire. And the best way to do this is to start off with some of the most interesting questions (Frey, 1983). That's why good surveys put questions about the person's background (education, age, occupation, religion, race/ethnicity, etc.) at the end. Items for the beginning of the survey not only should be interesting, but should not be on sensitive matters. Especially in interviews, it is important to allow a degree of trust and familiarity to build up before raising questions that may cause the respondent some discomfort—for example, questions about whether they ever drink more than they should or have recently been stopped for speeding.

It often is especially effective at the start of an interview or the beginning of a questionnaire to tell respondents what the study is about and why their participation is worthwhile. A good example of this, and of leading with more interesting questions, comes from a recent nationwide telephone survey conducted by Statistics Canada:

Hello, I'm_____ from Statistics Canada. We are conducting a survey in your area and throughout Canada on accidents and crime and their impact on Canadians.

All the information you provide will be kept strictly confidential. While your participation is voluntary, it is essential if the survey results are to be accurate.

These first questions ask about your opinions on crime and accidents and about ways in which people protect themselves and their property.

Compared to other areas in Canada, do you think your neighbourhood has a higher amount of crime, about the same or a lower amount of crime?

1.___ Higher
2.___ About the same
3.___ Lower
4.___ Don't know

Having begun the survey, it usually is best to group questions according to topic so that there is a coherent flow to the items. However, researchers often return to a topic several times, sometimes merely to see if the responses late in the interview (or questionnaire) match up with those asked earlier. And that brings us to the issue of response bias.

RESPONSE BIAS

We have explored how items can be biased. Now let's examine two biases commonly found in respondents and how these can be minimized.

CONFORMITY

Most people care about the opinions of others. Even when talking to an interviewer they have never met before and will not meet again, some people are more concerned about making a good impression—about conforming to social expectations—than giving the most accurate answer.

As it pertains to survey research, **conformity** refers to the tendency of respondents to select answers on the basis of their perceived social standards, to select "normal" or noncontroversial answers.

To offset this problem, it is necessary to make it as easy as possible to select answers that are not up to what might be perceived as ideal standards. Thus, for example, having asked people whether they ever drink alcoholic beverages, the General Social Survey does *not* ask them how often they get drunk. Instead, it asks them, "Do you sometimes drink more than you think you should?" It's much easier to say "Yes" to this question, and the same function is reasonably well served. The contrasting items about God, shown earlier, offer another example of how questions can be softened to make it easier for people to answer.

Another aspect of the problem of conformity is acquiescence. Some people want to be agreeable so they will tend to say "Yes" or to agree with little regard for the content of the question. For example, both of these questions were included in a survey (Lenski and Leggett, 1960):

It is hardly fair to bring children into the world, the way things look for the future.

Children born today have a wonderful future to look forward to.

Ten percent of the respondents agreed with both questions!

With reference to survey research, **acquiescence** is the tendency of some respondents to say "Yes" or "Agree" (some experts refer to this as yeah-saying).

One way to overcome this problem is to introduce a few questions with which virtually everyone would disagree early in the interview to get the respondent accustomed to disagreeing.

RESPONSE SET

Large batteries of items on a similar topic can be habit forming. Suppose you are answering a set of items about college life. All have the same set of response categories. As you begin to answer these questions, you find that all the early ones are favorable statements about college life and that you are answering each the same way. Growing impatient, you stop listening or reading carefully and keep on with the same response, failing to notice that the items toward the end say unfavorable things about college life. You have just fallen victim to response set.

Response set refers to the tendency of respondents to fall into a pattern of responding, having no regard for variations in content.

The solution is to alternate negative and positive items in order to prevent such a pattern from developing.

INTERVIEWING GUIDELINES

This is not a how-to-do-it manual for interviewing—whole books have been written on the subject. Nor is it the purpose of this course to prepare students to step right into an executive position in a survey organization. That said, a few basic points about the selection and training of interviewers are appropriate here.

SELECTION

Good interviewers are able to subordinate their own opinions and feelings in order to ask questions in a neutral way and to hear and record answers in an equally neutral, friendly way. Even if a respondent agrees that Hitler was a good man, interviewers must recognize that it is precisely their job to elicit that answer if that's what the respondent believes. Judgmental and argumentative people seldom are any good at interviewing. Interviewers also must be absolutely honest and conscientious. It is vital that they record answers faithfully and conduct interviews in the manner that the researchers intend.

It may be a good idea to seek interviewers who are somewhat similar to the intended respondents. The combinations of age, sex, race, or ethnicity of interviewers and respondents sometimes influence responses—especially if questions are asked on topics germane to the differences. For example, when questioned about race relations, respondents will tend to be a bit more open with an interviewer of their own race (Hatchett and Schuman, 1975; Weeks and Moore, 1981; Finkel, Gutterbock, and Borg, 1991). However, the more surprising result of studies of this phenomenon is how little it usually matters whether an older, white female or a young, African-American male is interviewing an older, white female or a young, African-American male. What seems to matter far more is whether an interviewer is a good listener and projects a warm and friendly impression.

TRAINING

Interviewers must be extremely familiar with their interview schedules. They also should practice interviews before they attempt to do any real interviews. Interviewers must be trained to read the questions precisely, always using the exact wording. The entire purpose of closed questions is to maximize comparability. If questions are reworded by various interviewers, comparability is severely compromised. The same applies to responses—any and all open-ended responses must be written down exactly as said and written legibly. Probes must be offered as provided in the interview schedules. Finally, interviewers must be helped to realize how important they are—that if they do a bad job, the survey is ruined.

NONRESPONSE

Chapter 4 discussed the problem of nonresponse in survey samples. Here we will discuss means for reducing nonresponse and then examine several ways for detecting the impact of nonresponse on findings.

MOTIVATING RESPONDENTS

Traditionally, surveys have relied on persuasion. For example, respondents might receive a preliminary letter or phone call explaining the importance of the study and seeking to convince them to participate. Questionnaires often come with a cover letter explaining why the participation of the respondent is extremely valuable. For example, a questionnaire study of police officers included a letter that began this way:

> *These days everyone talks about the police, but nobody listens to them. We plan to change that. And we can if you will tell us about your experiences in law enforcement and share your views on issues of special interest to police officers.*

Although people often can be persuaded to take part, today survey researchers have begun to offer people money, merchandise, and prizes to take part. Given the immense cost per interview, surveys can be conducted far less expensively by paying respondents a fee to fill out a questionnaire, even if the fee is substantial. The first questionnaire surveys to use monetary rewards offered trivial sums—some enclosed one or two one-dollar bills in the initial package. However, the idea was not to bribe respondents but to pose a moral dilemma for respondents. Should they send back the two dollars, which seemed like a lot of bother for so little? Should they keep the money, since they had not asked to be sent the questionnaire? Or should they complete the questionnaire, thus earning the money, even if it was only $2? Many people who otherwise would not have filled out the questionnaire took the third option, as this technique caused a significant rise in return rates (Yammarino, Skinner, and Childers, 1991). In addition, some market research phone surveys now offer merchandise or entry in a lottery for substantial prizes to respondents in return for being interviewed. Moreover, some of the leading survey organizations have begun to explore paying respondents for their time. Proposals to enclose a $25 or $50 certificate that can be redeemed for cash or merchandise upon return of a fully completed questionnaire are under review.

ASSESSING NONRESPONSE BIAS

There are several ways to try to discover how nonresponse may have influenced the results of a survey. Each involves an attempt to estimate differences between those who participated in the survey and the persons in the sample who did not take part.

The first method consists of obtaining some data on nonrespondents. For example, in an interview study, it often is possible to learn a few facts about respondents who eventually did not agree to be interviewed—their sex and perhaps other background factors such as age and occupation. Sometimes, brief telephone interviews are conducted with nonrespondents to obtain key information. Sometimes, a very brief mail questionnaire is sent. The cover letter might read,

> *Dear* _____
> *We realize that you were unable to take part in our recent survey on* _____
> _____. *Because it is vital that we be able to adjust our data to take account of those unable to take part, we would be very grateful if you would provide the information requested on this short questionnaire.*
> *Thank you.*

The questionnaire would ask age, education, race/ethnicity, and perhaps one or two questions on key topics. A crime victimization study might ask whether the respondent had been the victim of a crime in the past 12 months. A health survey might ask the respondent to rate his or her current health status. Such data would, for example, make it possible to ask whether the amount of victimization found by the survey was over- or underestimated by the failure to reach some respondents.

Another method assumes that nonresponse is not a categorical but a *continuous* phenomenon. That is, respondents differ along a continuum of willingness to participate, and this willingness varies considerably even among those who do respond. Thus, in questionnaire studies, each questionnaire is dated as it comes back on the assumption that the sooner people respond, the more willing they were to take part. Interview studies often carefully record how much effort was required to obtain the interview, and interviewers rate the attitude of each respondent. This information allows respondents to be ranked along a continuum of probability of responding. The key assumption is that *those lowest on the continuum tend to resemble those who did not, in fact, respond at all.*

Given this assumption, researchers check to see if variations in the probability of responding are related to other variables of interest. If, for example, in a crime victimization study, those least likely to respond do not differ from other respondents in terms of victimization, then it is likely that nonresponse is not influencing this measurement. Or, if those who were last to send back a health questionnaire are in as good health as early responders, then it seems reasonable to suppose that nonresponse did not influence the parameters found in the sample.

When these techniques do reveal nonresponse biases, it often is appropriate to correct for them by assigning *weights* to cases on the basis of groups undersampled or on differences in the probability to respond (see Chapter 4).

STUDYING CHANGE

Studies of change based on surveys take two forms: trend studies and panel or longitudinal studies.

Trend studies are based on two or more surveys conducted at different times, but based on independently selected samples of the *same population* and which include identical (or adequately comparable) questions.

Panel or **longitudinal studies** are based on surveys of the *same respondents* over a period of time, sometimes 10 years or more, to see how the respondents have changed. Panel studies usually span a rather short time, while longitudinal studies tend to cover relatively longer periods.

TREND STUDIES

Survey research has only recently come into its own as the basis for studying longer-term historical trends, because large-scale surveys based on probability samples are of rather recent origins. Many trend studies are based on the General Social Surveys of the United States, which now span more than 20 years and which carefully have repeated many questions many times. For example:

I am going to name some institutions in this country. As far as the people running these institutions are concerned, would you say you have a great deal of confidence, only some confidence, or hardly any confidence at all in them?

Congress	1973	1993
A great deal	24%	7%
Only some	61%	51%
Hardly any	15%	42%

The Press	1973	1993
A great deal	23%	11%
Only some	62%	50%
Hardly any	15%	39%

Science	1973	1993
A great deal	41%	41%
Only some	52%	52%
Hardly any	7%	7%

Clearly, these data substantiate a sharp decline in public confidence in Congress and the press, but confidence in science has held remarkably steady over this 20-year span.

PANEL OR LONGITUDINAL STUDIES

Surveys conducted at one point in time often are called **cross-sectional studies** because they depict a cross-section of the population at a particular moment. But some surveys are repeated with the same respondents— each interview or questionnaire being identified as a **wave** (the first wave refers to the initial data collection). Often, surveys are repeated within a relatively short period such as when three waves of interviews are conducted during the three or four months prior to an election. These short-term, repeated surveys are referred to as *panel studies*— political scientists who coined the term thought of the respondents as a panel of experts advising them on the course of a campaign. Sometimes, however, a sample of respondents is tracked over a very long period of time—as long as 10 or 20 years—and they are surveyed many times during this duration. These long-term studies, wherein respondents are surveyed repeatedly, are known as *longitudinal*[5] studies. (The names often are used interchangeably, and the distinction between the two may be unimportant.)

There are two very important differences between studies of change based on repeated, cross-sectional surveys of different respondents (but from the same population) and on those that reinterview the same respondents over time. The

---

[5] The dictionary defines longitudinal as "lengthwise"; hence, longitudinal studies follow people's lives lengthwise.

first of these has to do with what kind of change each measures, and the second has to do with nonresponse.

Trend studies are well suited for telling us *how a society or some other universe has or has not changed* over time. For example, by comparing the 1973 and the 1993 General Social Surveys, we can assert that support for capital punishment rose from 63 percent in favor in 1973 to 77 percent in favor in 1993. Each percentage is based on a national sample of the U.S. adult population drawn independently at a 20-year interval.

Panel studies also tell us about social change because, although the set of respondents remains the same, the period of time that passes is so short that these respondents remain representative of the population. Consequently, we can conclude from a panel study that candidate X steadily lost public support as the campaign wore on.

Longitudinal studies are not designed to measure social change. First of all, they seldom (if ever) are drawn from the general population but, rather, from some subset such as high school seniors or five-year-olds. And, while the sample continues to be representative of the population from which it was drawn, that population quickly ceases to be representative of the original subset. Suppose we have drawn a sample of high school seniors. In a year, they will no longer represent high school seniors but only the subset of persons who were in high school last year. Therefore, as we follow this sample over time, we can't really say we are studying how the population is changing but how a set of individuals is changing as it matures. That is, longitudinal studies chart how *a set of people change*, not necessarily how the world around them is changing. For example, it might be the case that people always change in these ways simply as a function of getting older and that, therefore, these changes are not in response to any changes in the environment.

All surveys suffer from nonresponse. But panel and longitudinal surveys suffer more as, each time the respondents are resurveyed, some cases are lost and, therefore, compared with the original sample, nonresponse increases. This cumulative loss of cases is referred to as *mortality*—some cases are, in fact, lost because respondents die between waves. Because of the much longer span of time involved, mortality is far more severe with longitudinal than with panel studies, and anyone undertaking such a study must be prepared to keep in frequent touch with each respondent in order to have his or her current address.

AGE, COHORT, AND PERIOD EFFECTS

Survey analysts must untangle three interwoven aspects of change.

Age effects occur because people change as they get older. For example, people may become more satisfied with their lives as they mature.

However, older people may differ from younger people, not because they changed as they got older, but because their generation always differed from later generations. These aspects of change are called **cohort effects**.

Cohorts are defined as persons within some subgroup of a population who share a significant life experience or event within a given period of time, usually from one to ten years. Usually, the shared experience is birth, and nearly all cohort analysis in social science is based on **birth cohorts**—people born within the same period of time (Glenn, 1977; Menard, 1991).

Finally, surveys may find substantial differences over time because *everyone* is changing. These differences are called **period effects** to indicate that they reflect the influences associated with a particular historical period.

Suppose we want to see if attitudes about race relations are changing. This item has appeared in most of the General Social Surveys:

Do you think there should be laws against marriages between blacks and whites?

First, we would want to examine age effects (based on the 1973 survey):

Age:	20–29	30–39	40–49	50–59	60–69	70–79
Yes, should be laws	21%	33%	40%	43%	48%	70%

Clearly, older people differ considerably from younger people. It seems rather unlikely that people are becoming more prejudiced as they age, so we would favor the cohort interpretation of the table—that each younger cohort has started out less prejudiced. On purely statistical groups, however, in this table we can't separate aging from cohort effects; *both* fit these data *equally well*.

Now let's take another approach and compare the results of three surveys spanning three *periods*.

	1973	1984	1993
Yes, should be laws	38%	25%	16%

Here we see that there has been a very substantial drop in the percentage who believe interracial marriages ought to be against the law. But is it a pure period effect in that *everyone* is changing? Or is it a pure cohort effect in that people aren't changing with the times, but that older, *more prejudiced cohorts are dying* and being replaced by less prejudiced cohorts? Both interpretations are entirely consistent with the table.

So, let's try something else—let's compare the same birth cohorts in different periods.

Percent: Yes, should be laws

	1973	1984	1993
20–29	21%	13%	11%
30–39	33%	13%	8%
40–49	40%	26%	13%
50–59	43%	35%	14%
60–69	48%	41%	27%
70–79	70%	54%	44%

In this table we can see all three effects. Reading down each column, we see that prejudice increases with age, although the pattern is inconsistent under age 60 in the 1993 data. Reading across each row of the table, we can see that, within all age groups, prejudice decreases with period. When we read down the table diagonally (guided by the sloping lines), we can see cohort effects—the cohort of people who were 20–29 in 1973 make up those who were 30–39 in 1984 and constitute the 40–49 group in 1993 and so on. Notice that each younger birth cohort is less prejudiced. But, when we examine the period effects within each cohort, we discover that they are limited to those cohorts made up of people who were under 40 in 1973. Prejudice does not decline across the three periods for the older cohorts.

Social scientists have devoted a great deal of statistical effort to untangling all three effects from one another (Glenn, 1977; Menard, 1991). This has proved to be very difficult because, as you can see in this table, no matter in which direction you read, you always are looking at a perfect mixture of two of these effects. For example, when you read across, you see both period and cohort effects. When you read down, you see age and cohort effects. And when you read down the diagonal, you see both age and period effects. Consequently, *only two* of these effects can be incorporated as independent variables in multivariate statistical analysis, such as regression analysis. *Any two* can be used, but not all three at once. This is referred to as the *identification problem* (Blalock, 1966), and it occurs when, in a set of independent variables, any one is a perfect function of one or more others. Recall from the discussion of using dummy variables for regression (in Chapter 5) that one must never use the dummy for each category of a variable at the same time, but only the total minus one. Thus, if you wish to introduce the categorical variable for region into a regression and it is coded into four dummy variables (one for each of the major regions), you will use only three as independent variables in an analysis since the fourth is perfectly predicted from the other three (anyone not living in the East, South, or Midwest necessarily lives in the West). Or, when using a dummy for gender, you may not use the dummy for percent male and the dummy for percent female at the same time, since each is a perfect predictor of the other. You will learn more about these kinds of limitations in a statistics course.

Data Archives

In a very important sense, survey data are never out of date. What would historians give for a survey of Rome in the time of Nero or of the United States on the eve of the Civil War? Future historians will have such data available for our era, and many historians of modern times already base their research on surveys conducted as far back as the 1930s. Moreover, many social scientists produce first-rate research, not by collecting new data, but by analyzing data already gathered by other researchers. Most surveys collect so much data that they are never "used up."

Reuse of surveys is possible because several archives exist to preserve surveys and to supply them to anyone who wants them. There are two primary survey data archives in the United States (and they contain a number of studies from other nations). The Roper Center for Public Opinion Research, at the University of Connecticut, maintains more than 10,000 surveys, including all those conducted by the Gallup Poll and by the Roper Poll, going back to the mid-1930s. The Inter-university Consortium for Political and Social Research, at the University of Michigan, is even larger than the Roper Center, although its studies tend to be of more recent vintage. Survey data files can be obtained from these sources for a modest fee. Both archives publish extensive catalogues of their holdings and welcome visitors.

Ethical Concerns

The committees who must give prior approval to all proposals for research involving human subjects use many rules. But, in the end, these rules come down to one simple principle: *In the conduct of social research, there must be no coercion, and no harm must come to those who are studied.* For survey researchers, this ethical principle has two primary implications.

First, an ethical researcher will always make it clear to respondents that they are entirely free to not participate and to refuse to answer any particular question or questions. This is especially true in potentially coercive situations such as college classes. Students must not feel required to fill out a questionnaire, for example.

Second, the privacy of respondents must be preserved. No person's answers ever may be revealed. For cross-sectional surveys, the privacy of respondents usually is preserved by anonymity. That is, once the interview is complete or the questionnaire returned, the name and address of the respondent is destroyed, making it impossible for anyone to link any interview or questionnaire with a specific individual. This anonymity is not possible for panel or longitudinal studies. The identity of respondents must be maintained until the last wave has been completed. This is because each new wave must be merged with prior waves, which requires that the researchers be able to link each respondent to his or her data file. Here the solution is a system to ensure confidentiality. Each respondent is assigned an identification number. Each new wave of data carries these identification numbers, thus allowing researchers to add to each respondent's file. However, the master list linking specific respondents to their ID numbers is kept extremely confidential, often in a safety deposit box, and is available only to the principal investigators. It is not considered ethical to tell respondents they will be anonymous when their answers will only be confidential.

Thus far, the courts have not acknowledged surveys as privileged information. On the other hand, no survey researcher ever has been forced to appear in court and reveal information about a respondent.

CONCLUSION

The aim of this chapter was to give you an overview of the practice of survey research. It should have prepared you to evaluate the many survey results that fill the media and to recognize typical flaws. As with the remaining chapters in the book, it also is meant to serve as a reference for planning and conducting an actual study—should your instructor make that part of your particular course. The workbook exercises will give you some hands-on experience with some basic survey techniques covered in the chapter.

REVIEW GLOSSARY

- A **simple fact** is an assertion about a concrete and limited state of affairs, often merely the claim that something happened or exists and usually having to do with only one or very few cases.

- A **proportional fact** asserts the distribution of something, or even the joint distribution of several things, among a number of cases.

- An **attitude** is an enduring, learned predisposition to respond in a consistent way to a particular stimulus or set of stimuli. We all have attitudes toward significant things in our environment: objects, people, ideas, events, and so on. An attitude usually is regarded as consisting of three components: (1) a belief, (2) a favorable or unfavorable evaluation, and (3) a behavioral disposition (DeLamater, 1992).

- **Behavior** involves action; it is what living creatures *do*. We "hold" attitudes, but we "perform" behavior.

- **Closed questions** force all respondents to select their responses from a set the researchers provide.

- **Open-ended questions** permit respondents to answer as they wish, in their own words.

- A **cafeteria question** allows respondents to pick and choose from a large selection of responses, indicating only the ones they select.

- A **forced option question** requires a response to every option.

- **Contingency questions** divert respondents to different questions or direct them to skip questions, *contingent* on responses to a prior question.

- As it pertains to survey research, **conformity** refers to the tendency of respondents to select answers on the basis of their perceived social standards, to select "normal" or noncontroversial answers.

- With reference to survey research, **acquiescence** is the tendency of some respondents to say "Yes" or "Agree" (some experts refer to this as yeah-saying).

- **Response set** refers to the tendency of respondents to fall into a pattern of responding, having no regard for variations in content.

- **Trend studies** are based on two or more surveys conducted at different times, but based on independently selected samples of the *same population* and which include identical (or adequately comparable) questions.

- **Panel** or **longitudinal studies** are based on surveys of the *same respondents* over a period of time, sometimes 10 years or more, to see how they have changed. Panel studies usually span a rather short time, while longitudinal studies tend to cover relatively longer periods.

- A **cross-sectional study** represents one point in time.

- Each instance of data collection from respondents in a panel or longitudinal survey is referred to as a **wave**.

- **Age effects** occur because people change as they get older. For example, people may become more satisfied with their lives as they mature.

- However, older people may differ from younger people, not because they changed as they got older, but because their generation always differed from later generations. These aspects of change are called **cohort effects**.

- **Cohorts** are defined as persons within some subgroup of a population who share a significant life experience or event within a given period of time, usually from one to ten years. Usually, the shared experience is birth, and nearly all cohort analysis in social science is based on **birth cohorts**—people born within the same period of time.

- Finally, surveys may find substantial differences over time because *everyone* is changing. These are called **period effects** to indicate that they reflect the influence associated with a particular historical period.

Comparative Research: Using Aggregate Units

Many of the most fundamental questions pursued by social scientists are not about individuals. Why is the homicide rate 10 times higher in Louisiana than in Iowa?" is not really a question about individual behavior. Rather, it directs our attention to aggregate units of analysis—in this instance, the 50 states. Viewed at the individual level, people in Louisiana and Iowa are fundamentally alike. But, viewed as aggregates, the states in which they reside are extremely different. To explain why, we must examine states, not people. Other questions would cause us to select high schools, baseball teams, or computer software companies as the units of analysis. These examples make it evident that there are two basic kinds of aggregate units.

Areal units consist of aggregate units having geographic boundaries—an area. Commonly used areal units are nations, states, counties, cities, and neighborhoods.

Social units consist of aggregate units having social boundaries—the basis of inclusion is social, not geographic. All persons residing within a clearly designated area belong to the aggregate making up Chicago. But only a few of these people belong to the aggregates making up the Chicago White Sox or Cubs.

As mentioned in Chapter 6, studies based on aggregate units usually are referred to as *comparative research*. As we shall see, some methodological issues are special to comparative research, among them the measurement properties associated with all aggregate units of analysis. Consider that a city may have an average family size of 2.3 children, but no family in that city has three-tenths of a child.

RATES AND BASES

To explore why homicide rates vary so greatly from one state to another, we would proceed by comparing states. However, as soon as we set about this task, it would be apparent that comparing states is different from comparing individuals. There is far more basic variation among the 50 states, for example, than among any 50 people. Consider that Alaska has an area of 656,424 square miles. In contrast, Rhode Island's area is only 1,545 square miles. On the other hand, there are only about 600,000 people in Alaska, or slightly fewer than 1 per square mile. "Little" Rhode Island is nearly twice the size of Alaska if population is made the basis of comparison, and the state has a population density of 960.3 people per square mile. Such variations are typical of most aggregate units of analysis. As a result, it is necessary to eliminate some of these variations in order to make meaningful comparisons. For example, it makes no sense to compare California and Wyoming on the basis of the 2,005 murders that took place in California and the 11 that happened in Wyoming in 1999. A state with a population of more than 32 million ought to have many more murders than a state with about 480,000 inhabitants. To make states comparable on such things as number of murders, we create rates.

A **rate** reduces raw numbers to a common base. Often, rates are based on population, but many other bases also are used to compare aggregate units.

The official crime statistics reported each year by the FBI consist of rates per 100,000 population. Thus, the number of murders is divided by the total population and the result is multiplied by 100,000. This calculation informs us that Louisiana had the highest murder rate, 10.7 per 100,000 inhabitants in 1999, Rhode Island's rate was 3.6, and California's rate was 6.0. New Hampshire and Iowa were tied at 1.5 for the lowest rate.

Frequently, however, population is not an appropriate base. Some rates, such as the amount of land under cultivation, are more appropriately based on geographic area than on population. Usually, we would prefer to know what percentage of a state, for example, is under cultivation than to know how many acres per 1,000 population are being farmed.

Even when rates should be based on population, many are improved greatly by being based on *only a relevant portion* of the population. Birth rates are an excellent example. The number of infants being born in particular aggregate units usually is reported in terms of two quite different rates: the crude birth rate and the fertility rate.

The **crude birth rate** is simply the number of births in a given year per 1,000 population.

The **fertility rate** is the number of births to the average female during her lifetime (this is estimated in a variety of ways).

The crude birth rate ignores the fact that only some members of the population can have babies. The fertility rate takes this fact into account. As it happens, the crude birth rate is as good a measure as the fertility rate when the age and sex profiles of the aggregate units are very similar. But, when units differ greatly in their proportions of female residents in their childbearing years, the two rates can give extremely different results. This is true in Canada. The Yukon Territory has a very high crude birth rate—second highest among the 12 provinces and territories of Canada. This would suggest that women in the Yukon have a lot of children. Not so. When the fertility rate is examined, the Yukon falls from second highest to next to lowest. Relatively many babies are born in the Yukon because it has an unusually high proportion of women in their childbearing years—57 percent as compared with 47 percent for the rest of Canada. But these women actually have relatively fewer children than do women in most other parts of Canada.

Abortion rates pose similar concerns about the proper base. Sometimes, the rates reported are based on the entire population, ignoring the fact that only pregnant women can have abortions. So public health statisticians compute a rate based on the number of live births. That is, the number of abortions per year is divided by the number of live births and the result is multiplied by 1,000. We can use these calculations to discover that recently the rate for New York State is 634 abortions for every 1,000 live births while in South Dakota there are only 82 abortions per 1,000 live births.

In both of these examples, it was reasonably easy to select the proper base for the rates. Often, however, it is not nearly so clear that there is a "right" base. Suppose you wanted to measure the modernization of agriculture of European nations. One of the variables available is the number of traction animals in each nation. Traction animals are used to pull things and include horses, mules, reindeer, oxen, and buffalo.[1] Converted into a rate, this ought to be a valid measure of modernization—where farmers still rely on traction animals, they are far less productive. One rate could be based on population, which would eliminate variations in the number of traction animals based solely on population size. Another rate might be the ratio of traction animals to tractors and trucks—an attempt to see what proportion of the pulling is still being done by animals. A third rate could attempt to eliminate variations in the amount of farming being done by basing the rate on the number of acres under cultivation. As it turns out, each rate produces somewhat different findings in that each classifies European nations in a somewhat different order.

As a final example, suppose you wish to measure the political climate of states, attempting to identify how liberal or conservative they are. One possibility is to examine the circulation of magazines devoted to political opinion. *The National Review* is a magazine of conservative opinion, and *The Nation* is a magazine of liberal opinion. So, for each state, you calculate the number of readers of *The National Review* per 100,000 population and do the same for *The Nation*. However, when you examine circulation rates for *The National Review*, the results defy common sense. Vermont, Massachusetts, New York, Oregon, and California come up as the most conservative states, while Mississippi, Alabama, and South Carolina come up as the most liberal. The rates for *The Nation* show that Connecticut, New Hampshire, Virginia, Vermont, and Massachusetts are the most liberal, while Kentucky, West Virginia, Arkansas, and Utah are the most conservative. In fact, the two rates are highly, positively correlated (0.61). The problem is that political magazines tend to sell better in the same states, regardless of content.

The problem can be eliminated by changing the base of the rates. Instead of using population as the base, we can divide one circulation rate by the other to see the relative popularity of each—this eliminates variations in the tendency to subscribe to magazines. If we calculate the number of readers of *The National Review* per reader of *The Nation*, the results become very plausible. In South Carolina, there are more than seven *National Review* readers per subscriber to *The Nation*. In Mississippi and Alabama, there are five. In only three states do readers of *The Nation* outnumber readers of *The National Review*, and they are Vermont, New York, and Oregon. These results would seem to have at least face validity.

RELIABILITY

It is often noted that aggregate data tend to produce far higher correlations than are found in analyses of survey data—something you have undoubtedly noticed while doing your workbook exercises. The reason for this is simple: Aggregate

---

[1] Not the "buffalo" of North America, which actually are bison.

data usually are far more reliable than are data based on individuals (Duncan, Cuzzort, and Duncan, 1961). Put another way, there usually is far less measurement error involved in aggregate measures. As a result, correlations are not reduced by random errors in classification, as they tend to be in survey data. Equally important, in the absence of significant random measurement errors, aggregate data do not produce small correlations when, in reality, there are none. Consequently, not only do we observe many very high correlations between variables based on aggregate data, we also see correlations of zero (while random fluctuation often makes surveys yield small, if meaningless, correlations).

BENEFITS OF LARGE NUMBERS

Aggregate measures tend to be more reliable simply because they are based on such large numbers of observations. Measurements based on individuals often depend on a single datum. For example, we may classify an individual as a political liberal on the basis of a single check mark on a questionnaire. But, when we classify Massachusetts as a state that tends to vote for liberals, the measurement is based on the actual votes cast by hundreds of thousands of people in many elections. The chances of misclassifying Massachusetts are infinitesimal, but the chances that any given individual will be misclassified are significant. By the same token, a delinquency rate for a city is based on records involving many thousands of people, while surveys construct a delinquency score for individuals on the basis of answers to several self-report items.

This discussion is in no way intended to denigrate survey data. As noted in the previous chapter, for certain purposes there are no better data to be had. But it is important, nonetheless, to recognize the superior reliability of aggregate data—other things being equal. Of course, other things aren't always equal and aggregate data can be unreliable, too.

LIMITS OF "OFFICIAL" STATISTICS

Most data available for aggregate units, especially areal units, are collected by governments. Immense amounts of aggregate data in all nations come from the censuses. But many other government agencies, at all levels, also "keep books" on social life—on crime, births, deaths, disease, marriage, divorce, suicide, education, manufacturing, retail sales, public expenditures, poverty, welfare, and much more. Government fact collectors also record an abundance of data on other socially relevant variables such as weather, pollution, food and water supplies, irrigated land, paved roads, and abundance of wildlife. However, official statistics are subject to certain limits that can decrease their reliability.

Omissions and Biases. Many official statistics have limits because of omissions and biases in the way they are defined or collected. Often, these shortcomings are unavoidable and well known but significant nonetheless. Official crime statistics offer an example.

The homicide data considered earlier in the chapter are gathered by local police agencies that, in turn, report them to the FBI. Justice Department statisticians collate all these reports and, once a year, publish the *Uniform Crime Report*, wherein crime rates are reported for all states, all Metropolitan Statistical Areas, all cities and towns having 10,000 or more inhabitants, and for many colleges and universities. In similar fashion, *Canadian Crime Statistics*, an annual volume published by the Canadian Centre for Justice Statistics, breaks down crime rates for the major areal units.

Both the Canadian and American crime statistics are calculated as rates per 100,000 population. Some of these rates suffer from unreliability related to using less than appropriate bases. All living persons are at risk of being murdered and, therefore, homicide rates based on the entire population would seem appropriate. But the same is not true for rape. In 1999, the official U.S. rape rate was 32.7 per 100,000 population. But, by definition, only rapes involving a female victim are included in the calculations. Consequently, the 1999 rape rate would rise to 70.0 per 100,000 if only females were included in the base. But are *all* females the proper base? Despite lurid media reports, children and older women seldom are raped. If the rate were adjusted to a base made up only of females 15 to 50, then the rate would be 130.0. Which rate is correct? None of them!

American and Canadian official crime statistics include *only those crimes known to the police*, either because they were reported or because the police discovered them. It is well known that many crimes are never reported. The best estimate is that only about 60 percent of all rapes are reported, so the true rate for the United States in 1999, based on females 15 to 50, might have been as high as 182.0 per 100,000.

Murder is by far the most fully reported crime. Of less serious crimes, only about 25 percent of larcenies are reported, while 75 percent auto thefts are reported—they must be reported if owners are to collect on their insurance. In any event, the official crime rates systematically underreport crime.

We know much about underreporting because, for the past several decades, the U.S. Department of Justice Statistics has conducted an annual victimization survey, which asks all members (at least 12 years old) of 49,000 households about any crimes of which they were victims during the previous year. While the victimization survey uncovers many unreported crimes, it too has biases (Jacob, 1985). First, the survey discovers a crime only if it discovers the victim (or victims). No murder victims are interviewed. The survey doesn't even try to discover gambling or prostitution offenses since the "victims" also are offenders. Second, it greatly underreports crimes suffered by persons under age 12, since they are not interviewed. Third, it is susceptible to underreporting as well as to overreporting— some respondents forget or are unwilling to report victimizations, and others make up things to impress interviewers.

Now for the good news. The flaws in the official crime rates matter only if we are attempting to estimate exactly how much crime is taking place. But criminologists are satisfied that the rates are sufficiently reliable to allow accurate comparisons across states, cities, and counties (Blumstein, Cohen, and Rosenfeld, 1991). While the true crime rates are undoubtedly higher for each areal unit, the official crime rates are thought to be sufficiently well correlated with the true rates so that units are ranked or ordered properly on each variable. That is, North Carolina's true burglary rate in 1999 most certainly was higher than the official rate of 1,286.9 per 100,000, just as New Hampshire's true rate was higher than its official rate of 307.9. But it is very likely that North Carolina had the highest true rate and New Hampshire the lowest. By the same token, British Columbia's true rate probably was well above the official rate of 2,143 per 100,000, but it very likely had the highest true rate among Canada's 10 provinces, and New Brunswick very likely had the lowest true rate just as it had the lowest official rate.

This illustrates an important measurement principle: *Measures that may not be sufficiently accurate for descriptive purposes may be entirely adequate for explanatory analysis.*

This rule applies not only to aggregate data. Recall the two items from Chapter 7 asking about government spending. One question found that a majority thought too much was spent on "welfare" while the other question found that a majority thought too little was spent on "assistance to the poor." Clearly, each question yields a somewhat inaccurate description of how the American public feels about efforts to mitigate poverty. But each gives identical results when used as a dependent variable—recall the discussion of item interchangeability from Chapter 5. We return to this matter in the section on the validity of aggregate measures.

The point of this discussion of crime rates is to stress the fact that, just because numbers come from official sources and appear in thick books, their reliability is not guaranteed. Indeed, in some parts of the world, numbers from official sources are particularly suspect. Until the reforms that led to its breakup, the Soviet Union had crime statistics (and many other statistics) that were known to be extremely underreported. The official line was that crime is virtually nonexistent in socialist nations, and the crime statistics were "edited" to conform to this doctrine. In similar fashion, in 1994, a commission appointed by the Chinese government reported widespread and extensive falsifications in official statistics to the extent that such basic facts as the actual fertility rate or the amount of rice production could only be guessed at within rather wide limits. A fairly obvious tip-off that statistics are fabricated is when they don't change. For example, some developing nations report an unemployment rate of 2 percent year after year. Infant mortality rates are suspiciously static in some nations, too. And so it goes.

As these examples make clear, when aggregate units are international, as when nations are the units of analysis, problems of omission and bias often are magnified. If official U.S. crime rates depend upon the accuracy of reports from more than 16,000 separate city, county, and state law enforcement agencies, consider the probable unreliability of crime statistics collected in nations where local record keeping often is a hit-or-miss affair and where reports to higher authorities are often fabricated. According to the crime statistics published by

INTERPOL[2], in 1990 China reported an official rape rate of 4.3 per 100,000 and a burglary rate of 17.5. Guatemala's official rape rate was put at 4.11 and its burglary rate was 27.18. Not only are these rates implausibly low, they raise another issue: Rates carried to several decimal places suggest a degree of precision that is absurd when the actual procedures by which the data were collected are considered.

False Precision.   Every month, the press reports the official unemployment rate. A typical report reveals that unemployment rose or fell by one-tenth or two-tenths of a percent—from 5.9 percent to 6.0 percent, for example. Despite the fact that changes of this magnitude inspire many solemn comments about the direction in which the economy is going, and despite the fact that the stock market often responds to these announcements, the official employment rate is far too unreliable to be taken that seriously. Shifts of less than one full percentage point (and possibly even two points) probably are too small to have any meaning, and the rate itself is a rather dubious measure of what would sensibly be meant by the concept of unemployment (Mandel, 1994). In a subsequent section, we will return to this matter as we consider the validity of aggregate measures. Here the point is that a significant portion of the variation in unemployment rates across states and cities may be nothing but measurement error—variations that are within the confidence limits of the statistics (see Chapter 4).

False precision becomes an even more serious matter when international data must be relied upon. The United Nations Food and Agriculture Organization (FAO) publishes estimates of world grain production, basing its conclusions on data assumed to be accurate to 0.0001 percent. Such claims of precision are absurd. It would be impossible to measure the grain production of even one American farm anywhere close to that degree of accuracy. Assume that a very tiny farm produced exactly 21,312 pounds of rice. To achieve the accuracy assumed by FAO statisticians, the measurement of this rice crop could not be off by more than plus or minus 2 pounds (or precisely 2.1312 pounds). No farmers would ever bother trying to measure a crop so precisely. And, when reporting the size of their harvests, farmers would be apt to use even less precise figures than they actually measured—"Well, I guess I got about 10 tons," our hypothetical farmer might tell a government agent.

Nevertheless, the FAO bases its estimates of per capita food supplies on these production estimates and issues very precise figures about the number of malnourished people in the world—especially the number in developing nations. A recent study looked behind these numbers.

Nicholas Eberstadt (1995:171) cited FAO statistics showing that Chad's food supply rose by precisely 0.3 percent between 1977 and 1980 and that, in 1980, per capita food supplies in Afghanistan and Chad differed by 0.4 percent. Eberstadt pointed out:

---

[2] The International Criminal Police Organization, headquartered in France.

. . . for the periods in question, however, it is thought that upwards of 90 percent of the populace in both countries was rural and illiterate, and that as much as half the production of goods and services in both countries may have occurred in the nonmonetized economy. . . . [Moreover] both nations were in the throes of chaos in 1980, convulsed by turmoil attendant to foreign invasion and civil war. No authority in the world could hope to measure a change in national diet of scarcely a pound of grain per person per year under those conditions, as the FAO was claiming to do.

Again, these defects do not mean that variations in per capita food availability cannot be studied in a set of aggregate units made up of nations. It does mean that, at the very least, all such rates should be scrutinized as to plausible levels of precision and many should be rounded off to a much lower level of precision before being used. However, as will be discussed later, many published rates for aggregate units are very accurate and justify the level of precision attributed to them.

VALIDITY

The validity of measures based on aggregate data must be assessed in the same way as other kinds of social scientific data are—do the indicators really measure the concepts they are intended to measure?

As with other indicators used by social scientists, the primary test of validity is simple face validity. Church membership rates have compelling face validity as measures of religiousness, and sales of hunting and fishing licenses seem valid measures of an outdoors lifestyle. But the validity of various other indicators is more problematic and in need of testing. One way to test validity is to look more closely at how the indicator is defined and calculated. When we look more carefully at an indicator, we may discover that its claim to face validity is superficial and that it really isn't measuring what we thought it did.

FAULTY INDICATORS

Recall President Coolidge's famous tautology: "When more and more men are thrown out of work, unemployment results." What he probably was trying to say may be closer to what most of us mean by unemployment than what the official rate actually measures. That is, unemployment means people *being out of work*. But the official rate measures only the number of people who currently are *actively seeking work*. People who lie and say they have looked for jobs in the past week are counted as unemployed, even though they may have no wish for work. But people who have stopped looking because they have looked too long unsuccessfully are not counted. That people often stop looking for work helps to explain the seeming oddity that the official unemployment rate usually rises when the economy is booming and often falls when there is a recession. When there appear to be more jobs available, many people (especially women and students) are motivated to look for work and by doing so are counted as being in the labor force and as unemployed. Moreover, not only are many people who truly are unemployed not counted in the rate, but, as noted, others are counted who perhaps ought not be. Anyone who is receiving unemployment benefits is counted. But it is well known that some workers plan their periods of unemployment the same way they plan vacations and do not seek work until their benefits are nearly used up. In addition, every spring the unemployment roles swell as new high school and college graduates seek their first jobs. Most don't consider themselves unemployed, at least not during their first several weeks of looking, but the Labor Department does.

Given this situation, it appears that the official unemployment rate is invalid as an indicator of a concept defined as the percentage of people who are out of work. There are two solutions available (in addition to simply dropping interest in the concept). The first is to redefine the concept to reflect what the official rate measures. If that is undesirable, the second is to seek more valid measures. Several years ago, some economists began to publish a new rate based on the number of column inches of help-wanted ads appearing in a national sample of newspapers. Their claim was that, when the volume of help-wanted ads declines, unemployment is higher and, when the volume increases, unemployment is lower—that the volume of advertising reflects the supply of persons available to be hired. This measure has not attracted much attention from either the media or the financial community. However, one thing seems clear: Whatever the help-wanted index measures is not the same as what the official unemployment rates measure. This is obvious in that the two measures are essentially uncorrelated. That brings us to another primary method for assessing validity.

ASSESSING VALIDITY

Earlier in this chapter, we examined the construction of a measure of a state's political climate on the basis of the ratio of readers of *The National Review* to readers of *The Nation*. As mentioned, this measure seemed to have face validity and appeared to rank states in a plausible order. But it is possible to apply a more stringent test of the validity of this measure. Recall that *convergent* validity involves demonstrating that a particular measure or indicator relates to other measures or indicators believed to measure the same concept. Consequently, if we can create some other measures of a state's political climate, each having substantial face validity, then each ought to be highly correlated with our original measure. For example, our measure of conservatism ought to be highly correlated with the percent voting Republican in recent presidential elections (omitting 1992 because Perot's third party muddled the results too greatly). It also ought to be negatively correlated with the percentage of women in the state legislature and the rate of volunteers for the Peace Corps over the past several decades. As it turns out, these predictions are strongly confirmed.

Many aggregate variables can be tested for validity in this fashion. For example, measures of infant and of childhood mortality not only ought to be extremely highly intercorrelated, but each should also be highly correlated with rates of death in childbirth, overall life expectancy, and availability of medical services. In similar fashion, measures of economic development—per capita income, percent employed in manufacturing, electricity consumption, telephones per 1,000, percent literate, etc.—ought to be highly correlated, if they are valid. In fact, such correlations have been found repeatedly.

To sum up: Although official statistics can suffer from omissions and biases, the fact is that *most of the available aggregate statistics are very reliable and valid*. This is especially the case for the more developed and democratic nations—not only in terms of national-level data, but also for subunits within these societies: cities, counties, states, provinces, prefectures, and the like. For example, data on number and type of motor vehicles are extremely accurate since they always are based on registration records. Or commonly used data on the composition of populations—the age and gender distribution, racial and ethnic makeup, percentage of female-headed families, or percentage of college graduates—come from each nation's census and can be relied upon for all of the developed nations and for many of those that are less developed. Indeed, variations across societies in the reliability of their census data can play a major role in determining which cases social scientists select for study.

SELECTING CASES

In assembling a set of aggregate units, selection should be guided by two major concerns: selecting cases that have adequate *comparability* and obtaining a sufficient *number* of cases to warrant statistical analysis. The latter concern often will determine whether the cases are sampled from a larger universe or whether all cases meeting the selection criteria are included.

COMPARABILITY

It is quite true that all humans are much more alike than they are different. Nevertheless, it wouldn't make sense to take national surveys of the United States and Japan and merge them into one data file. Differences in culture and in many basic social structures would be uncontrolled, and many distributions based on the combined samples would be very unlike those in *either* nation, as we see in this example based on the World Values Surveys conducted in 1990:

Overall, how satisfied or dissatisfied are you with your home life?

	United States	Japan	Total
Very satisfied	59%	17%	44%
Somewhat satisfied	28%	45%	34%
Dissatisfied	13%	38%	22%

Noncomparable Measures of Infant Mortality

✦   ✦   ✦   ✦   ✦

The press frequently reports that the infant mortality rate—the number of infant deaths during the first year of life per 1,000 live births—is substantially higher in the United States than in many other nations. The *Statistical Abstract of the United States* (1994) reported the U.S. rate to be 8.1, while Canada and Spain had a rate of 6.9, Germany and Switzerland had a rate of 6.5, Hong Kong's rate was 5.8, and Japan's was 4.3.

The blame for the elevated American rate usually is placed on inadequate prenatal health care for poor women. This may indeed be a contributing factor, but a substantial amount of the difference between the U.S. rate and that of many other nations is the result of differences in the way the rate is measured.

The World Health Organization's guidelines require that all births having any sign of life must be included among the number of live births, regardless of the size of the newborn or the length of the pregnancy. This standard is followed by the United States as is demonstrated by the fact that each year thousands of American infants weighing less than 2.5 pounds are included in the statistics on live births. These are, of course, extremely premature babies who have a very high risk of death—some years they make up nearly a third of all infant deaths reported in the U.S. mortality statistics.

Unfortunately, some nations exclude most premature babies and otherwise compromise the WHO standards in various ways. In Switzerland, for example, babies must be at least 12 inches long at birth in order to be counted as living, regardless of their vital signs. This excludes nearly all babies weighing less than 2.5 pounds and greatly reduces the Swiss infant mortality rate. In similar fashion, although Spain does not indicate that it excludes high-risk, small babies from its statistics, it reports only about 15 percent as many low-birthweight babies as do other western European nations. In Canada and the United States, more than a third of all infant deaths occur within the first day of birth. In some other nations, including France and Hong Kong, very few first-day deaths are reported, suggesting that many of their local reporting units do not count these as live births (Eberstadt, 1995).

Reliable measures require cross-case comparability. When noncomparable standards are applied, as they are in the case of infant mortality rates, it becomes impossible to adequately test hypotheses about international variations in infant mortality. One possible solution is to limit analysis to cases using the same measurement standards.

As can be seen, neither the United States nor Japan much resembles the total column of this table. But, if merging these two data sets makes no sense, it would make good sense to use the World Values Surveys—which asked the same questions of national samples in 42 nations—to create an aggregate data set made up of nations and use this data set to test hypotheses about why people in some nations are more satisfied with their home life.[3] Just as it would be inappropriate to merge an American and a Japanese survey, it would be equally inappropriate to mix aggregate units representing cities and nations. Aggregate data sets should consist of comparable units—units having enough in common so that comparisons are meaningful. Nations have rural areas, cities do not. Cities have retail districts, but many neighborhoods within cities do not. These are examples of incomparable units of analysis. One can compare nations, cities, or neighborhoods but not mix these units. The same principle applies to social aggregates. It would be appropriate to compare high schools, colleges, grocery stores, or TV stations but not to mix them.

NUMBER

Statistical analysis techniques require a substantial number of cases in order to produce meaningful results. Earlier in the chapter, we noted that, because aggregate data are based on so many internal observations, they yield rates of immense stability and reliability compared with survey data, and consequently they tend to produce very high correlations. Also because of this far greater reliability, aggregate data analysis requires far fewer cases than does survey analysis. This is fortunate because, although the world contains billions of people, each of whom might be surveyed, the world offers only a very limited number of aggregate units. For example, according to the current count by the U.S. Bureau of the Census, there are only 192 nations on earth. Not all of these can be included in a data set of nations, since there are almost no data available for some of them. Consequently, most nation data sets are limited to 100 cases or fewer.

Within nations, the number of cases is limited, too. There are fewer than 300 Metropolitan Statistical Areas in the United States and only 25 in Canada. There are only 50 United States, 55 counties in England and Wales, 47 prefectures in Japan, 30 provinces in China, and 77 neighborhoods in Chicago and 49 in St. Louis. As it turns out, there are enough cases in each of these sets to sustain solid research. However, it would make no sense to reduce the number in any of these sets by sampling cases. Nevertheless, some aggregate data sets are based on samples. For example, there are 3,141 counties in the United States. Sometimes, researchers work with samples of these counties, rather than use all of them. More often, they restrict their analysis to a subset such as all urban counties or all southern counties. Limiting the number of counties is especially common when researchers plan to code new variables for each case and not rely entirely on variables already available for all counties. By drawing a sample, researchers reduce the amount of time required to code variables.

---

[3] Within that data set, satisfaction with home life is highly correlated with the percentage who say their spouse shares their attitude about sex and who agree that both spouses ought to share household chores.

When social rather than areal units are the basis of research, the number of available cases also varies greatly. Thus, there are more than 3,000 American colleges and universities and more than 1,500 separate religious bodies in the United States. On the other hand, there are only 50 state legislatures and 31 National Football League teams.

ANALYZING AGGREGATE DATA

For more than 150 years, social scientists have been doing comparative research. In the 1820s, André Michel Guerry analyzed data for the 89 departments of France—calculating rates for suicide, homicide, illegitimacy, taxes, and military desertion. His book, *Essai sur la statistique morale de la France*,[4] published in 1833 by the French Royal Academy of Science, is thought to be the first attempt to analyze aggregate data. Soon after, other scholars followed suit, publishing studies based on such sets of aggregate units as the counties of England and Wales, the cantons of Switzerland, and the many states and principalities that now make up Germany.

THE CASE-ORIENTED APPROACH

Despite this early beginning, this was not the way comparative studies usually were done. Instead, most scholars utilized what often is called the *case-oriented approach* to distinguish it from the *variable-oriented approach*. The **case-oriented approach** selects two or more (but rarely more than five) cases and examines them closely in order to explain some striking difference or differences between (or among) them.

The list of those taking this approach includes the most distinguished scholars of the nineteenth century, among them Alexis de Tocqueville, Max Weber, Emile Durkheim, and Karl Marx. The tradition flourishes today in the work of such scholars as Barrington Moore (1966), Immanuel Wallerstein (1974), Theda Skocpol (1979), and Charles Tilly (1984). Case-oriented studies comparing Japan and the United States have generated a virtual publishing industry.

The case-oriented comparative method often produces work of impressive scholarship, bursting with insights and ideas. Indeed, such works are and continue to be fertile *sources* of concepts and theories. But the method is *unsuitable for testing hypotheses*. Consequently, Neil Smelser (1976:157) refers to it as "systematic comparative illustration" in that, with only several cases, it is impossible to perform any empirical operations other than to "illustrate" cause-and-effect statements. For example, it is true, as we just saw, that satisfaction with home life is far lower in Japan than in the United States and it also is well known that the average American family also has far more living space in their home than does the average Japanese family. But that is not a correlation. It merely illustrates the possibility that crowding causes family conflict, which results in dissatisfaction, for example. If variations across two cases were accepted as correlations, then *all*

---

[4] *An Essay on the Moral Statistics of France.*

ways in which Japan and the United States differ would be correlated—indeed, these would be perfect correlations (1.0).

THE VARIABLE-ORIENTED APPROACH

Theories can be tested only by checking out their empirical implications on an appropriate set of units of analysis—by seeing whether correlations predicted and prohibited by the theory meet expectations. To do this requires enough cases to produce stable statistical results. Thus, the focus shifts from specific cases to variation across a set of cases—to a variable-oriented approach.

The **variable-oriented approach** tests hypotheses by applying statistical techniques such as correlation and regression to variables based on an appropriate set of aggregate cases.

Even when using an adequate set of cases, aggregate data present certain methodological challenges. One of these is, of course, the *ecological fallacy*, which was discussed in detail in Chapter 3. Suppose we found a high correlation between satisfaction with home life and amount of living space, using a substantial number of nations as the units of analysis. We could not be sure that this correlation would hold at the individual level. Aggregate units are not people, and it is possible statistically that ecological or aggregate correlations do not reflect individual-level behavior—although there is a considerable literature on statistical procedures by which inferences about individual-level correlations can be drawn from aggregate-level correlations (Langbein and Lichtman, 1978).

Levels of Measurement.   In aggregate data sets, variables below the level of interval measures are rare. In fact, nearly all aggregate variables meet the criteria of ratio measurement. Crime rates, fertility rates, magazine circulation rates, per capita income, illiteracy, and percent employed in agriculture all have meaningful zero points and equal intervals across values. Therefore, most aggregate data fully meet the assumptions of sophisticated statistical procedures such as regression analysis. This is fortunate in that the number of cases available for analysis often is quite small, precluding cross-tabular analysis.

Discovering Outliers.   Although aggregate variables usually have meaningful zero points, they often have no ceiling or at least a very high one. When survey respondents are asked how happy they are, they cannot exceed the highest of the categories offered them. But there is no limit on per capita income, and although in principle a homicide rate cannot exceed 100,000 per 100,000 (wherein everyone has been murdered), this limit leaves room for extraordinary amounts of variation. Combine immense variation with relatively small numbers of cases, and a serious statistical problem arises.

An **extreme outlier** is a case having such a deviant value on a variable that it distorts correlations, sometimes causing an apparent correlation where none exists among the other cases and sometimes suppressing a correlation that does exist among the other cases.

Suppose that, in a survey data set including 1,600 cases, a clerical error assigned one person a value of 87 on the happiness question while all other cases had values of 0 (very unhappy) to 4 (very happy). Suppose that a researcher then correlated this variable with another measuring income. The misclassified case would have no detectable effect on the correlation. But suppose that there were only 50 cases in the data set. This case might cause a correlation all by itself.

In aggregate data sets, it does not require clerical errors to produce a case having a value on some variable that is extremely high or low compared with other cases. This can have extreme consequences for statistical results. As a result, good analysts use scatterplots to examine each correlation of interest for the possibility that it is being distorted by an outlying case or cases. Let's pursue an example.

Using the 50 states as the units of analysis, we discover that there is a very strong and highly significant correlation between the marriage rate and the rate of alcohol consumption: 0.56. No matter how long we ponder this coefficient, we can't discover whether or not it is a good description of the cases. To do that we need to actually see the joint distribution of the cases on these variables. A scatterplot lets us do this (you will already be familiar with scatterplots from your workbook exercises). Here is the scatterplot:

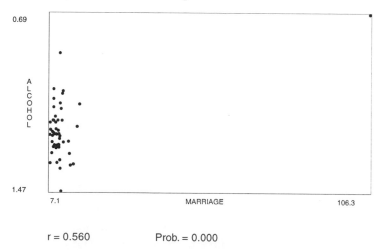

r = 0.560   Prob. = 0.000

Notice that all but one of the cases are tightly bunched at the left side of the scatterplot, while in the far, upper right of the screen there is a single case. It turns out that this case is Nevada, the state with the highest alcohol consumption rate and with an incredibly high marriage rate. Its annual rate is 106.3 per 1,000 population, while Hawaii, the next highest state, has a rate of only 16.2. This suggests that more than 10 percent of the population of Nevada (including infants) marries every year! Nevertheless, Nevada's rate is not in error. It is generated by its immense tourist-marriage industry. That is, thousands of tourists come to Nevada every year to be married in one of the state's many marriage chapels and to then honeymoon at a casino hotel. Consequently, Nevada's marriage rate is spectacular. It also distorts correlations.

Any well-designed statistical analysis program not only will allow you to examine scatterplots, but also will include an automatic function for locating outliers and removing them. With Nevada removed, the scatterplot looks like this. The correlation has collapsed to an insignificant –0.037. That is, there is no correlation between these variables in the other 49 cases.

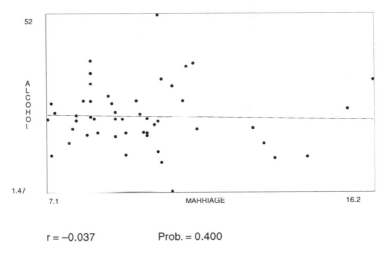

r = –0.037   Prob. = 0.400

When analysts discover an outlier such as this, they should exclude it from correlations involving the variable or variables on which the case is an outlier. An appropriate and efficient way to do this is to substitute the missing data code for the value of that case on the variable.

Significance and Censuses?   You will notice that significance was reported for each of these scatterplots. But, these findings are not based on a sample. The 50 U.S. States are a universe or population and, when all 50 are included, it is a census, not a sample. In Chapter 4, we defined a *test of significance* as a calculation of the odds that a correlation is produced by random fluctuations between the sample and the population. When all cases are included, no such random fluctuations can occur—what we observe is the actual parameter, not the statistic estimating that parameter. So why report significance?

Statisticians often have advised that significance is irrelevant when the data are based on a census and, therefore, should not be calculated or reported (Morrison and Henkel, 1970). If the only objective of research is to generalize to a specific population, then this view of significance is correct. However, if the objective is to test hypotheses, then this view of significance is too restricted (Blalock, 1979).

Suppose we discover that there is a clear regional pattern in the distribution of public libraries per 10,000 population so that the states of the upper Midwest have the highest rates. Then, as we prepare to formulate hypotheses to explain this regional effect, we encounter critics who say this finding is merely chance. To demonstrate their point, these critics use random numbers to assign values to each state, creating 100 such randomly generated variables. Then they report that five of these random variables display a regional effect as marked as the one produced by the library rate. Clearly, then, the possibility exists that *any* regional effects *could* be random.[5]

Hubert Blalock's (1979:242) response to such a claim cannot be improved upon:

Notice that there is no question, here, of generalizing to a population larger than the total of 50 states. The argument revolves around the processes that could have generated differences among subpopulations delineated in various ways. Certainly, if one could have frequently obtained differences as great as the regional differences by using a table of random numbers, then . . . [social scientists] should always make tests in order to rule out the simple "chance-processes" alternative.

The probability of chance processes producing correlations is especially high when very small numbers of cases are the basis of the analysis. Since this is typically the case in comparative research, there is sufficient justification for always using tests of significance.

Statistical Limits of a Small n. In statistics, *n* stands for the number of cases on which any given calculation is based. As noted, aggregate data sets often consist of relatively small numbers of cases. Since all tests of significance take *n* into account, no bias results. However, the smaller the *n*, the fewer variables that may be used in any given empirical model such as regression analysis. In principle, no model may include more variables than the number of cases minus one,[6] for to do so would exceed the degrees of freedom (you will learn about this in a statistics course). In practice, that standard is much too permissive. While there is no mathematical basis for it, experienced researchers suggest that it probably is unwise ever to include more than one independent variable per eight to ten cases. Using that standard, analyses based on the 50 states are limited to five to eight independent variables.

STUDYING CHANGE

Studies of social change necessarily involve aggregate units and nearly always areal units. Sometimes, these studies are comparative and variable-oriented as when researchers attempt to see the effects of foreign investment on economic development in Third World nations, comparing changes in 62 nations from 1965 through 1988 (Firebaugh and Beck, 1994). Others are case-oriented comparisons—for example, comparisons of the changes that led to revolutions in France, Russia,

---

[5] This would not include regional patterns for which the cause is clearly established historically such as the large proportion of African Americans in southern states or the high levels of population density in the East.

[6] This limit differs depending on the statistical technique being used.

and China (Skocpol, 1979). Some track changes in one society over time and employ a statistical method known as *time series analysis*. But, whatever their methodological approach, all studies of change depend upon the existence of historical data.

HISTORICAL DATA

It might seem likely that quantitative studies of social change would be limited to very recent times. But that's not really the case. A wealth of historical data exist, if not always in ready-to-use form. A few examples will give some sense of what's already available. Then we will see how social scientists have succeeded in coding variables from surviving records and other sources of information.

Available Data.   Governments have been gathering and reporting interesting and reliable statistics for about 200 years. France began to gather national statistics on crime as long ago as the end of the eighteenth century. The results showed that the numbers were incredibly stable from one year to the next, although differing greatly from one city to another. Another striking finding was that gender differences also were extremely stable. Thus in 1826, 21 percent of those arrested for theft were women while in 1830 women made up 22 percent. Men and women also differed greatly in terms of the means they used to murder someone. From 1825 through 1830, women committed only 11 percent of murders involving violent methods, but committed almost half of murders by poison.

As discussed in Chapter 4, the first census of the United States was conducted in 1790. By the middle of the nineteenth century, the census had been greatly expanded in scope, and an immense variety of information can be found in old census volumes. For example, beginning in 1850, census enumerators asked whether anyone in a household had died in the past year and, if so, what was the cause of death. These data are not useful as an overall picture of causes of mortality during this period—too often, the precise cause of death was not known to the survivors ("old age," "he just took sick and died"). But some of the causes were quite clear and offer data of great interest. For example, the census of 1860 reported that 984 persons had been reported murdered during the previous year (79 of them women). If we turn that number into a rate, we can estimate that there were at least 3 murders per 100,000 population, compared to about 9 per 100,000 today. The 1860 rate is certainly too low. Victims who did not leave behind a surviving household to report their deaths to census enumerators would not have been counted. Since these are precisely the kinds of people—male drifters without families—most apt to become murder victims, the rate is underreported. So, comparisons with today are tenuous. But comparisons across the states and territories and from 1860 to a later decade elude the problem. Consider this portrait of murder in 1860:

Murders per 100,000, 1860

New Mexico Territory	155.6	North Carolina	2.3
Utah Territory	32.3	Virginia	2.1
Texas	20.0	Illinois	2.1
California	19.7	Rhode Island	1.7
Kansas	14.9	Massachusetts	1.5
Oregon Territory	13.3	Ohio	1.4
Washington Territory	9.0	Iowa	1.3
Nebraska Territory	7.0	Maryland	1.3
Louisiana	6.9	Connecticut	1.3
Arkansas	6.4	Indiana	1.3
Florida	6.4	Michigan	1.2
Kentucky	4.7	South Carolina	1.1
Missouri	4.1	New Jersey	1.0
Georgia	3.7	Pennsylvania	0.9
Alabama	3.6	New York	0.9
Tennessee	3.6	Maine	0.8
Delaware	3.6	New Hampshire	0.6
Minnesota	3.5	Wisconsin	0.5
Mississippi	3.4	Vermont	0.0

Obviously, the Old West was much as it has been depicted in the movies. Even though the rates are considerably underreported, the homicide rates for the various territories and for Texas, California, and Kansas are high in comparison to modern rates—the New Mexico Territory was an especially dangerous place. Rates in the settled East were much lower—there were no murders at all in Vermont during the year.

There is nothing unique about American data. As mentioned, the French began to collect data on many topics during the 1820s. Soon after that, the bureaucracy in Paris was demanding an incredible array of statistics from local officials and publishing them annually in huge volumes—all of which survive. By the 1870s, these many volumes were summarized in a large annual volume, *L'Annuaire Statistic de la France*, which continues to be published. The nineteenth-century volumes were reprinted during the 1960s and can be found in many U.S. and Canadian libraries, especially at the larger universities. Anyone wanting to know how many people were being sentenced year by year to Devil's Island (and for what offenses), how many suicides occurred, how many children were born out of wedlock, or how many Catholic priests were ordained can find it all here.

Similar troves of ready-to-use data exist for the other nations of Western Europe.

Creating Data.   **Charles Tilly** is one of the pioneers of quantitative studies of social change and particularly changes involving political conflict and violence. In an early paper, he tested ideas about the social basis for the revolt against the revolutionary French government that broke out in the Vendée in 1793. What Tilly had discovered was that an immense variety of good records survived, waiting only to be coded. Working with records of enlistments, executions, and the like, Tilly (1969) created a profile of the pro- and anti-government factions, finding merchants, small manufacturers, and peasant landowners (as opposed to farm laborers) to be very overrepresented among the anti-government faction. Ever since, Tilly has relentlessly searched out records shedding light on European rebellions during the nineteenth and early twentieth centuries, an effort that eventually also involved his wife **Louise Tilly** and his brother **Richard Tilly**. The basis for their influential study *The Rebellious Century, 1830–1930* (1974) was many variables from available sources such as the *L'Annuaire Statistic de la France*. But the central data on the frequency and severity of rebellions by year were created by the Tillys themselves.

First, they defined the events of interest: conflicts in which at least one group of 50 or more persons, acting together, seized or damaged persons or objects not belonging to themselves. Then, they sat down and searched daily newspapers having national circulations, recording information about every event that fulfilled their definition of a violent outburst. Having identified these events, the Tillys then searched elsewhere for additional information—police records, pamphlets, and historical works. In the end, they were able to trace the variations in French political violence rather precisely. Because the population of France grew substantially during this period, and because it seemed possible that a larger population has the potential to produce more violent outbursts, these raw numbers were converted to rates per million population.

An interesting feature of this research is that it is based on a single aggregate unit—France. Clearly then, nations are not the units of analysis in this study. The units consist of years. That is, each year has its number of violent events, and analysis consists of correlating violence with other variables across this set of years. This form of analysis is known as *time series analysis*.

TIME SERIES ANALYSIS

Many studies of change, particularly those conducted by economists, are based on repeated measures of the same variables over time. These are called *time series studies*.

Time series analysis examines the relationship between two or more variables based on the same universe but measured at a number of points in time. Therefore, in time series analysis, *points in time* are the *units* of analysis.

When economists want to see whether changes in interest rates influence home sales, for example, they collect data on both variables over time. They may use weeks, months, or even years as their units of analysis. To test the hypothesis that housing sales are negatively correlated with interest rates (that sales fall when rates rise), they would calculate the correlation between the two variables. If they wanted to include a second independent variable such as the unemployment rate in their analysis, they would use regression to discover the net effects of each variable (beta) and their joint effects (R^2). There are two basic forms of times series analysis: nonlagged and lagged.

Nonlagged Time Series.   When the independent and dependent variables are measured at the same point in time for any given case, this is a **nonlagged time series**.

American newspaper circulation rates have been falling for more than 20 years. That is, people are becoming less likely to read newspapers. Suppose you wanted to explain why. Many have suggested that TV is the cause—that, as people spend more and more time watching television, they read less and less. The General Social Surveys frequently have asked national samples of Americans how often they read a newspaper and how much time they spend watching TV. Your first step would be to create a data set in which years were the cases. Next, you would calculate for each case the percentage who read the newspaper daily and the percentage who watch a lot of TV. Notice that, although the data are based on surveys, individuals are no longer the units of analysis. Instead, the responses of everyone in the sample have been aggregated to characterize years. *People* either read a newspaper daily or they don't—only an *aggregate* can have a percentage of daily readers.

Were you to examine the variation of each measure across time periods, you would observe the expected decline in daily newspaper readership—from 66 percent in 1975 down to 46 percent in 1993. But there has been no corresponding increase in television viewing—none at all. Since TV viewing did not vary during this period, it cannot explain variations in newspaper reading—so much for the hypothesis.

Why, then, have people become increasingly less likely to read newspapers? Could it be that they have lost confidence in them as a source of information? Recall from Chapter 7 that, through the years, the GSS often has asked people how much confidence they had in various social institutions and that confidence in the press has declined substantially over time. To test the hypothesis that people have stopped reading the paper because they have lost confidence in the press, all you need to do is add this variable to each case representing a year. Figure 8.1 graphs these two variables over time—the curves are very similar. This is reflected in the correlation coefficient, which is .796 and significant at the .001 level. The hypothesis is supported.

Lagged Time Series.   Whenever the values of the independent variables or the dependent variables assigned to specific cases come from different time periods—when there is a systematic time lag among variables—this is a **lagged time series**.

Hypotheses often will specify a time lag between a change in one variable or set of variables and another. For example, economists don't anticipate that housing sales will fall within an hour after a rise in interest rates. Consequently, they may lag their measure of housing sales and calculate the correlation between interest rates at time 1 with housing sales at time 2 and so on. The lag might be a week, two weeks, a month, six weeks, or some other lag thought to be appropriate. Alternatively, it might be interesting to test the hypothesis that, in fact, rising home sales increase the demand for loans and thereby cause interest rates to rise. To test this hypothesis, economists would lag interest rates to follow home sales by some period of time.

Figure 8.1   Correlating Newspaper Reading and Confidence in the Press, 1975–1993

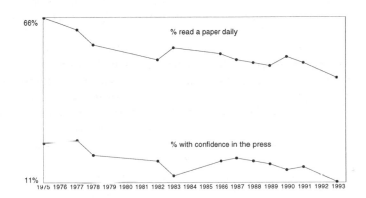

In this way, lagged time series let social scientists attempt to sort out questions of cause-and-effect—of causal order. How do analysts conclude which is the better model—which hypothesis about causal order is better supported? They pick the one that fits the data better—the one with the higher correlation in the case of a two-variable hypothesis and the one with the larger R^2 when three or more variables are involved (Ostrom, 1990).

This aspect of time series analysis is appropriate to the example concerning newspaper reading and confidence in the press because two alternative causal orders are plausible. Is it the case that people have stopped reading papers because they no longer trust them, or is it the case that they no longer trust papers because they have stopped reading them? To find out, we can lag each variable and compare the results.

When confidence in the press is lagged by one time period, the correlation falls to .655. In contrast, when newspaper readership is lagged, the correlation increases to .873. Although both correlations are significant at the .001 level, the substantial difference between the two coefficients tells us that the model postulating that the decline in newspaper reading caused a decline in confidence in the press fits the data much *less* well than does the model postulating loss of confidence as the cause of the decline in readership. Since this model also fits the data somewhat better than does the unlagged model, we should prefer it of the three.

Although all the leading analysis systems will allow analyses of time series data, most social scientists who do a lot of time series analysis use special programs designed exclusively for this purpose. These programs provide automatic lagging functions and will generate very detailed graphic displays of trends.

ETHICAL CONCERNS

Since so much aggregate data is taken from published sources, it might seem that research on aggregate units does not raise ethical concerns. No individuals suffer or have their privacy violated when it is known that Idaho leads the nation in hunting licenses or even that Michigan is the state with the largest percentage of overweight residents. But people can suffer harm when aggregate units are smaller. Even with areal aggregate units, problems of anonymity can arise. Consider a small town with only one Hispanic family. If the U.S. Bureau of the Census were to publish the average family income of different racial and ethnic groups in this town, anyone who cared to could find out what income the Hispanic family reported on its census form since the Hispanic average income would refer to that family alone. Thus, the smaller the unit, the more data the U.S. Bureau of the Census refuses to publish.

Problems of anonymity can become even more acute when social units are used. Suppose a study were based on 200 liberal arts colleges with enrollments of 800 to 1,500. Suppose, too, that some of the data came from surveys of the students at each school and that it included rates of drunkenness, cocaine use, sexual activity in dorm rooms, cheating on exams, and other potentially embarrassing activities. In such cases the researchers would have to be very careful to make sure that the results never became available in a form allowing specific schools to be identified. The press must never report that Muddletown College is "Number One" in drunkenness or that 78 percent of students at Doltsville College cheat on tests.

To prevent unethical dispersal of information, it isn't sufficient simply to omit school names from the data set. It may well be necessary to refuse to divulge the names of schools included in the sample or even the state in which they are located—it doesn't take many clues to figure out which case is which school if one knows what schools are included and something about the schools. Precautions are especially important if the data are to be placed in an archive and made available to other researchers.

As the number of social units declines, problems of identity are even more difficult. A study involving 20 sororities at a particular school might have a substantial potential for embarrassing individuals, depending on the information collected. This potential is, of course, apt to be considerably less than when field methods are used, for, as we see in the next chapter, fieldworkers often can't help but become aware of sensitive information.

CONCLUSION

This chapter is meant to acquaint you with the key aspects of research based on aggregate units. However, because even the most sophisticated aggregate analysts so typically base their work on public sources, it seems fitting to conclude this chapter by identifying sources to which you will have relatively easy access.

Data Sources

Many of these sources are in most libraries, and most of them are very inexpensive.

United States

From the U.S. Government Printing Office:

The Statistical Abstract of the United States. Published annually.

The State and Metropolitan Area Data Book. Published every several years.

The County and City Data Book. Published periodically.

Crime in the United States: Uniform Crime Reports. Published annually.

Sourcebook of Criminal Justice Statistics. Published annually.

Highway Statistics. Published annually.

Agricultural Statistics. Published annually.

Census of Retail Trade. Published every five years.

Census of Wholesale Trade. Published every five years.

Census Summary Volumes. Published every ten years.

From elsewhere:

Bluebook. Published annually by the Audit Bureau of Circulation, Schaumburg, IL. Contains circulation data on hundreds of magazines.

Churches and Church Membership in the United States. Published every ten years by the Glenmary Research Center, Atlanta, GA.

Canada

From Statistics Canada:

Census Summary Volumes. Published every five years.

Canadian Crime Statistics. Published annually.

Bluebook. Published annually by the Audit Bureau of Circulation, Schaumburg, IL. Contains circulation data on many Canadian magazines as well as circulation of many American magazines in Canada, broken down by provinces.

INTERNATIONAL

From the World Bank:

Social Indicators of Development. Published annually.

World Debt Tables. Published annually.

World Development Report. Published annually.

World Tables. Published annually.

From elsewhere:

The World Almanac and Book of Facts. Published annually. Sold in most super-markets.

Demographic Yearbook. Published annually by the United Nations.

Statistical Yearbook. Published annually by the United Nations.

Worldwide Military Expenditures and Related Data. Published annually by the U.S. Arms Control and Disarmament Agency.

Factbook. Published annually by the Central Intelligence Agency. Sold in most large bookstores.

The State of Food and Agriculture. Published annually by the U.N. Food and Agriculture Organization.

International Crime Statistics. Published annually by the International Criminal Police Organization.

World Health Statistics Annual. Published by the World Health Organization.

The Statistical Abstract of the United States. Published annually. (Last section devoted to international data.)

REVIEW GLOSSARY

- **Areal units** consist of aggregate units having geographic boundaries—an area.

- **Social units** consist of aggregate units having social boundaries—the basis of inclusion is social, not geographic. All persons residing within a clearly designated area belong to the aggregate making up Chicago. But only a few of these people belong to the aggregates making up the Chicago White Sox or Cubs.

- A **rate** reduces raw numbers to a common base. Often rates are based on population, but many other bases also are used to compare aggregate units.

- The **crude birth rate** is simply the number of births in a given year per 1,000 population.

- The **fertility rate** is the number of births to the average female during her lifetime (this is estimated in a variety of ways).

- The **case-oriented approach** selects two or more (but rarely more than five) cases and examines them closely in order to explain some striking difference or differences between (or among) them.

- The **variable-oriented approach** tests hypotheses by applying statistical techniques such as correlation and regression to variables based on an appropriate set of aggregate cases.

- An **extreme outlier** is a case having such a deviant value on a variable that it distorts correlations, sometimes causing an apparent correlation where none exists among the other cases and sometimes suppressing a correlation that does exist among the other cases.

- **Time series analysis** examines the relationship between two or more variables based on the same universe but measured at a number of points in time. Therefore, in time series analysis, *points in time* are the *units* of analysis.

- When the independent and dependent variables are measured at the same point in time for any given case, this is a **nonlagged time series**.

- Whenever the values of the independent variables or the dependent variables assigned to specific cases come from different time periods—when there is a systematic time lag among variables—this is a **lagged time series**.

CHAPTER # 9

Field Research

- **RELIABILITY**
  - DECEPTION
  - FIELD SURVEYS AND CENSUSES
- **VALIDITY**
- **SELECTING TOPICS AND SITES**
  - RELEVANCE
  - ACCESSIBILITY
  - RISK
- **ENTERING THE FIELD**
  - NEGOTIATING ACCESS
  - NEUTRALIZING OBSERVER EFFECTS
- **OBSERVATION**
  - UNSTRUCTURED
  - STRUCTURED
- **INTERVIEWING IN THE FIELD**
  - INFORMANTS AND RESPONDENTS
  - SYSTEMATIC OR IMPROMPTU INTERVIEWING
- **FIELD NOTES**
  - ORGANIZING FIELD NOTES
  - ANALYZING FIELD NOTES
- **STUDYING CHANGE**
- **ETHICAL CONCERNS**
- **CONCLUSION**
- **REVIEW GLOSSARY**

189

Surveys and censuses are the appropriate research methods for obtaining proportional facts. But neither is well suited to gathering data on *processes*, especially processes involving *interpersonal interactions*. For example, a survey is superior for discovering what proportion of women in a peasant village marry before age 18, but the survey method is ill suited to discovering the patterns of courtship that result in marriage. Surveys also are not well suited to learning the details about specific events—about what went on at the most recent wedding reception, for example.

To study processes and events, social scientists ought to go out and observe the phenomena of interest. That is, they ought to do their research in the field—in the actual settings in which the things they wish to study naturally occur. That is precisely why John Lofland and Rodney Stark went out in the field to study conversion to a new religious movement. For even interviews of people who have recently converted may not reveal the key elements involved in their conversions because people often fail to notice or recall accurately what went on. In fact, interviews conducted with converts almost always place primary emphasis on beliefs, on how the doctrines of the group attracted them to join. But, as they observed actual conversions take place, Lofland and Stark discovered that it was attachment to group members that played the primary role and that people did not even fully understand many doctrines until after, often long after, their conversions. The *only* way this could have been discovered was by direct observation of actual behavior as it took place—that is, by doing field research.

Field researchers have developed some special methods which we will examine in this chapter. But, in a very important sense, the field is a *place*, not a method. Therefore, some of the research techniques examined in other chapters can, and often should, be used in the field in conjunction with those methods more specific to field research. We will deal with these issues as the chapter proceeds.

Methods textbooks typically claim that one advantage of field research is that it doesn't require much funding. It is true that many field studies have been conducted by one person equipped only with a notebook and a pencil. But even these studies took a great deal of time to complete, during which the researcher needed some means of support. Moreover, as we see in the next section, the quality of field research is greatly improved if it involves multiple observers devoting full-time to the research. Thus, one of the most famous and productive field studies ever done involved a team of observers including Howard S. Becker, Blanche Geer, Anselm Strauss, and Everett C. Hughes, who spent more than two years observing students in one medical school and several more years writing up their results. While this may have cost less than a national survey, it was not cheap.

RELIABILITY

When properly conducted and applied appropriately, the results of field research are extremely reliable. As noted, field research is not well suited to supplying proportional facts, but it is unsurpassed for providing data on what people actually do in real life. Consider the following examples.

Richard T. LaPiere (1934) wrote a letter to a number of hotels and motels in California asking whether or not they would be willing to accept Chinese guests. Since a substantial number replied that they would not, had this been a survey study, the conclusion would have been that discrimination against the Chinese was widely practiced by California hotels and motels. But LaPiere didn't stop there. Instead, he took a Chinese couple to all the hotels and motels and discovered that the great majority of those whose managers had said they would not rent to Chinese did so without comment when confronted with real people. Obviously the field observations were a far more reliable guide to actual behavior.

As their study of conversion to the Unification Church (the Moonies) progressed, **John Lofland** and **Rodney Stark** (1965) discovered that they needed to determine who, among those they were observing, was "a convert." That is, some people said they were members and would have given that answer to a survey interviewer, but Lofland and Stark became suspicious that their professions of faith were insincere. So, Lofland and Stark devised some observational "tests." For example, they had observed that it was the norm that members of the group close their eyes during prayer. So, Lofland and Stark ceased doing so and instead looked to see if anyone else was breaking the norm. They immediately discovered that the two men whose sincerity was most suspect always looked around during prayers and would exchange "knowing" grins with each other and with Lofland and Stark.

In both of these examples, the greater reliability of field research is based on the ability to directly observe behavior rather than trying to infer how people will act on the basis of their verbal reports. However, even field researchers must rely on what people tell them and, in addition, we often must rely on what field researchers claim to have seen or to have been told. Consequently, as with all research methods, there are some common sources of unreliability which field researchers need to overcome.

There are several reasons for the lack of reliability in field research, all of them correctable. So, as we deal with these matters do not become disenchanted with what can be a very powerful research method.

DECEPTION

Field research usually is conducted by one researcher. That means there is no one available to help the researcher evaluate what he or she is seeing and hearing. Further, there is no one in a position to contradict anything the researcher claims to have seen or heard. Unfortunately, in many ways, field research resembles wildcatting for oil. Most expeditions into the field produce the equivalent of dry holes, but the possibility of a gusher motivates the continued search. Unlike oil drillers, however, fieldworkers can hit a gusher simply by saying that they did. The result has been that some of the most famous field studies eventually turned out to be, at the very least, extreme instances of wishful thinking—Margaret Mead's famous study of Samoa being an acutely embarrassing example (Freeman, 1983). The sad fact is that had Mead come back from Samoa with an accurate ethnographic report, it would not have made her famous.

Obviously, any specific field researcher can avoid this source of unreliability simply by being honest. It might also help to make field research more reliable if it became a far more collaborative effort involving multiple observers. Recall from Chapter 6 that Lofland and Stark's field study of conversion has been confirmed more than 25 times by subsequent researchers. Not only did this study have two primary field researchers, but Lofland and Stark managed to persuade several other sociologists to observe the group from time to time.

It isn't only field researchers who may engage in deception. Often, they are themselves deceived by those whom they are observing. Sometimes, the deception is intentional—people alter their behavior whenever the observer is around. For example, during Napoleon Chagnon's first season of observing the Yanomamö tribe in a remote part of the Brazilian rain forest, he saw them only as peaceful gardeners. It was only after repeated visits that he began to discover they weren't peaceful at all but were among the most murderous people ever studied (Chagnon, 1988).

However, often the people being observed deceive the fieldworker unintentionally. That is, people may give honest answers to the field researcher, but these answers may be wrong. This is especially likely when the answers involve proportional facts.

FIELD SURVEYS AND CENSUSES

In his total of 13 field trips to study the Yanomamö, **Napoleon Chagnon** did not observe even one murder, let alone an ambush involving multiple homicides. He knew about these murders only because some members of the tribe told him such things happened. But, even so, he had no awareness of the bloody reality of life among the Yanomamö until he decided to construct a genealogy for each of the tribe's 12 villages. To do this, he conducted a census, obtaining data on everyone including immediate family members who were dead. These data revealed that nearly 70 percent of adults over age 40 had at least one close relative—parent, spouse, sibling, or child—who had been killed by another member of the tribe. More than 30 percent of all male deaths were murders. Nearly half of the males over age 25 had taken part in a murder.

Now, when Chagnon says this is a violent society, he can give specific answers to the question "How violent?" But he can do this *only* because he conducted a census; no Yanomamö could tell him how frequently people took part in murders, because none of them knew. Without collecting systematic data from everyone, or from a representative sample, no one *could* know.

Unfortunately, in the past, field researchers often seemed allergic to numbers. Authors of books on how to do field research often included the most extraordinary polemics against numbers—Lofland (1976) even included a lengthy list of the character flaws that lead people to do quantitative social research. But even the most committed disciples of nonquantitative field methods constantly make proportional claims. Their publications are filled with words such as *typical, rare, widespread, unusual, unpopular, infrequent, many, most, few,* or *none.* All such words postulate a proportional fact. And, unless the group is extremely tiny, it is virtually impossible to determine proportional facts about it unless one systematically collects information on the variable or variables in question either from all members of the group (a census) or from a representative sample of members (a survey).

There is nothing about doing a census or a survey in the field that interferes with observational techniques. Consequently, it is becoming more common for field researchers to combine their traditional methods with systematic, quantitative techniques as well.

Lofland and Stark (1965:863) did not need a formal census or survey of the Moonies in order to report that "the converts were primarily white, Protestant, and young (typically below 35); some had college training, and most were Americans of lower-middle class and small town origins." It was easy to correctly typify the group because only 21 people were involved. But, a decade later, when **Eileen Barker**, a sociologist at the London School of Economics, began her study of the Moonies in Britain, there were hundreds of local members, and, during her period of observation, more than a thousand potential converts turned up at recruitment meetings. It would have been irresponsible of her to have compared the backgrounds of British Moonies with those reported by Lofland and Stark only on the basis of observations as informal as theirs.

Moreover, had she reported that "many" or "few" converts had received psychiatric help prior to their contact with the movement, or that "most" or "hardly any" had been very happy at the time they encountered the Moonies, she would have been criticized severely by other social scientists. These are "facts" that can't be known just by hanging around with a group, watching, and talking to some members. To think otherwise is to accept impressions and intuitions as satisfactory sources of facts.

Barker (1984) did not report her impressions and intuitions on these matters; instead, she collected systematic data. Because Barker's study is a prize-winning example of field research done right, we shall draw on it for many examples in this chapter. Here let's quickly see how she collected questionnaire data from people, contrasted those who converted with those who did not, and compared converts who remained members for several years with those who quit soon after joining.

In 1979, as she began her field research, Barker was able to get permission to distribute a questionnaire to all persons who attended one of the group's two-day workshops. These people were having their first exposure to the Moonie teachings. Eventually Barker obtained data on 1,017 people. In addition to collecting data from each newcomer on his or her first visit to the group, Barker was able to track their subsequent contacts. These data allowed her to chart the eventual recruitment outcomes of each of these potential converts. She discovered that, of those who attended an initial, two-day workshop, only 10 percent ended up joining the church for at least a week and only 5 percent remained for at least 2 years. These solid numbers dealt a fatal blow to claims that the Moonies used mind control or brainwashing techniques to gain converts. If this was brainwashing, it certainly didn't work very well.

Barker also was able to present important data on the mental health of those in her field survey. Keep in mind that these data were collected from people as they arrived to attend their first two-day workshop. Hence, these data were collected *before* anyone had converted and are not subject to retrospective distortions. Barker's data showed that the group of attenders as a whole were not a bunch of neurotic misfits and malcontents. Very few had been in psychiatric treatment, very few reported periods of depression, and hardly any were unemployed.

Even more important was Barker's finding that people who joined and stayed at least two years were substantially less likely to have been depressed, unhappy, unemployed, or to have had a history of psychiatric treatment than were those who didn't join or who joined but soon quit.

It is possible that had Barker simply associated with Moonies for several years, she would have come to suspect that it was the more normal and well-adjusted converts who stayed. But this would have been a hunch or an opinion that her peers would not have found convincing (most social scientists would have expected quite the opposite). If you want to make credible claims about the joint distribution of two variables in a population, there is no alternative to systematic data collection.

Eileen Barker adjusts an earring as she watches a Moonie meeting.

Of course, there were many things about life among the Moonies that were of great interest to Barker and which could not have been studied efficiently or effectively by means of a questionnaire. Her book, *The Making of a Moonie (1984),* mainly consists of careful descriptions of processes such as: How did members locate people and convince them to attend a workshop? How did they "court" potential converts? How did they neutralize defections so as not to let them undermine member morale? How did they adjust to an arranged marriage? Or, in Barker's (1984:16) own words:

> *What is life like in the Unification Church? What kinds of communication system and power structure does the organization have? To what extent, and why, does the movement vary according to time and place? What is the range of relationships the Church and its members have with the rest of society?*

To learn these things, Barker watched and listened.

VALIDITY

Validity involves actually measuring what you think you are measuring. Questions of validity are obvious for survey research. They are far less obvious for field research because rarely do researchers provide much information about how they classified their observations. Of course, simply ignoring the issue of validity doesn't make it go away, but field research methodologists have written very little on validity.

In general, however, field research has tended to be so descriptive as to evade validity issues. That is, to the extent that abstract concepts are utilized, they tend to have been arrived at inductively—as names applied to sets of empirical observations after the fact. Thus, for example, the field researcher may invoke the concept of role conflict after having observed chronic acrimony between waiters and cooks or between police and probation officers. Moreover, because so much field research is exploratory, it is not directed by theories and does not involve links between concepts and indicators. Thus, it is reliability, not validity, that is the primary concern.

SELECTING TOPICS AND SITES

All research begins by defining a topic of interest and identifying an appropriate set of units of analysis—what do we want to know and where can we look to find out? No matter what research method is to be utilized, there always are serious intellectual, practical, and ethical limits on the options. These can be especially acute and compelling for field research.

RELEVANCE

Since most field research is exploratory, the selection of things to study is not shaped by theoretical imperatives. This allows for immense latitude in selecting topics to study and places in which to study them. But with such freedom comes the burden of responsibility not to select a topic and/or site on frivolous, self-indulgent grounds. Far too many field studies were conducted for no better reason than that the researcher already was involved in a particular activity at a particular place. Thus, someone studied cocktail waitresses in an upscale hotel bar simply because she was one. Someone else studied a New Age discussion group because he was a deeply committed member. Someone else did a study of race track gamblers because betting on the horses was his favorite pastime. And a lot of people have studied various campus groups because they belonged to them. Sometimes, studies done because of such private involvements contribute to our fund of social scientific knowledge. Often they do not, being highly partisan "reports" disguised as objective research.

Another potential source of frivolous research might be identified as answering the call of the exotic. There have been many studies of strippers, prostitutes, transsexuals, cultists, nuns, nudists, survivalists, bikers, tattoo artists, carnival "freaks," and rock musicians. There are comparatively few field studies of receptionists, check-out clerks, dentists, retail clerks, day-care operators, Methodists, or auto technicians—far more significant groups socially, but lacking in glamour. While reports about exotic groups often are interesting, it is not clear that they add anything significant to the social scientific enterprise.

The abundance of field studies done for other than social scientific reasons has clouded the scientific standing of the method. That is unfortunate because field research is the only appropriate method for pursuing many very central social scientific questions.

Granted that good field researchers often go into the field without hypotheses, but they have clearly defined reasons for going, reasons that involve scientific relevance. That is, they have general research questions in mind. Lofland and Stark didn't have a hypothesis about how conversion occurred, but they knew they wanted to watch people convert in hopes of finding out why they did. That aim guided their search for an appropriate group and gave an initial focus to their observations.

Stephen Richer (1984) wanted to learn when and how children's play activities begin to reflect sex role differentiation. This led him to observe children at day-care centers.

Elliot Liebow (1967) was aware that studies of child-rearing practices among poor African-American, inner-city families were failing to include adult males. So he picked out a street corner where unemployed men "hung out" and began to observe.

People who begin field research with no specific questions to guide them must discover a reason during their research. This poses immense pressure on the researcher to see something interesting—even if it isn't there.

ACCESSIBILITY

Suppose a social scientist wants to study patterns of informal interaction between corporate officers and directors during board meetings of *Fortune* 500 companies. Or, what about a sociologist who wants to observe campaigning during the selection of a new pope? Forget it! There simply are many aspects of life that are not open to outsiders.

Sometimes, as in these examples, inaccessibility is obvious. But often it is not. For example, one would expect, given the social stigmas imposed on them, that unusual religious groups would be inaccessible. This is usually not the case. The Moonies have been extremely cooperative with many field researchers, for example, and so have many other similar groups. In contrast, many of the largest and most conventional religious denominations reject all requests from social researchers.

Thus, one of the first things field researchers need to discover is accessibility. Can they get in to do their study? Must they pretend to be members in order to gain access? We shall return to this issue, but, for now, we will consider a very central factor that intersects with accessibility: risk.

RISK

The real-world nature of field research means that many potential studies would involve serious risk (Lee, 1995). Joining a terrorist group in order to see how they recruit, how they control their members, or how they evade detection might produce valuable social scientific knowledge, but it would be exceedingly dangerous. Indeed, field research on illegal groups involves not only the risk of the observer being detected by the group, but also the risk of criminal liability as an accessory or as a co-conspirator. If the people you are observing commit a crime, you have no legal right to not report them.

Even when a researcher undertakes to study people or groups who are not themselves dangerous or criminal, substantial risks can arise. Recall from Chapter 4 that, when NORC sent interviewers to search blocks in order to interview the homeless, each interviewing team was accompanied by an armed cop. A field researcher who decided to hang out with the homeless for six months would have no such protection.

ENTERING THE FIELD

In Chapter 6, we contrasted the overt and covert observer roles. Covert observation may solve problems of accessibility—people who would refuse permission to be studied, get studied anyway. But it raises serious ethical problems and places severe limits on the observer's freedom to seek information. We are not prepared to say that covert observation is *never* ethically acceptable—clearly it is acceptable to make systematic observations of what people do in public places, and it would be an impossible requirement to ask an observer to obtain permission from each person passing through. But, when it comes to less public behavior, the informed consent of the observed becomes a matter of concern. Thus, if field researchers wish to discover what Moonies do and say in their churches and homes or what teachers do in their classrooms, they should have permission to conduct their study. Typically, that involves convincing persons in authority to authorize the fieldworker to proceed. This may take time, for the researcher not only must establish a basis for trust, but often must explain just what it means to do social scientific research. Because Eileen Barker wrote such an informative account of her negotiations with the Moonies, it will be an instructive example.

NEGOTIATING ACCESS

Barker's initial contact with the Unification Church came when she was invited by them to a conference for social scientists. It "turned out to be disappointingly respectable," but it soon led to another invitation from the Moonies to visit a farm they operated near London. This led to other, less formal contacts with some church members. In the process she "was becoming fascinated with the Unification Church."

At the time, the Moonies were the targets of very antagonistic depictions in the media, and this made Barker (1984:14) doubt that she would be able to conduct a study of them

> *unless I were to pretend to become a member myself. This was out of the question for a number of reasons. First, I would have been unhappy about the deception on purely ethical grounds; secondly, I had no desire to give up my job* [Moonie converts were required to devote themselves full-time to the movement]; *and thirdly, even if I were to have joined, I would not have been able to go around asking questions on any sort of systematic basis without arousing suspicion. I dropped a few hints that I would be interested in doing a study, but I did not hold out much hope.*

> *Then events took an unexpected turn. I was surprised one day to learn from my secretary that a Moonie about whom I had become rather interested (he had obtained a good history degree from Cambridge and I knew his father slightly) had been to visit me. . . . I was afraid he might get into trouble if I tried phoning, so I waited to see if he would make further contact. About a week later he found me in my office.*

The purpose of the visit was to propose that Barker write a paper about the Moonies as a response to a negative report being written by a sociologist whose information came entirely from people antagonistic to the group.

> *I explained that although I would be interested, I could not write a paper without first conducting a proper sociological study. I would need to have independent funding for my expenses, and I would require a complete list of membership in Britain so I could interview on a random sample basis, rather than seeing only those members whom the movement wanted me to interview.* (p. 14)

> *It took several weeks to convince the British leadership that I could not do the study without the complete membership list. I tried to reassure them that as a*

sociologist I had no interest in, or intention of, divulging information about identifiable individuals. . . . In the meantime I applied to the Social Science Research Council [for funding]. . . .

It was at the end of 1976, more than two years after I had first made contact with the movement, that I finally got permission to do the study on my own terms, with no conditions except for the promise, which I gave freely, that I would not publish personal details which could be related to identifiable individuals—except, of course, in the case of information which was publicly available and/or related to the known leaders of the movement. The Moonie who had come to see me told me some time later that the reason why they had been prepared to let me do the study was not because they thought that I would necessarily support them—they did not, he admitted, really know how I regarded the movement—but because I had been prepared to listen to their side of the argument, and they could not believe that anyone who did that could write anything worse than was already being published by people who had not come to find out for themselves. The final "test" came on New Year's Eve, 1976, when one of the American leaders came to my house with two British Moonies. We discussed the research for about an hour and he then left, apparently satisfied. My children spent several exciting but unrewarded hours searching for the "bugs" which they felt quite certain would have been planted in my study. (p. 15)

Thus, Barker began her field research in January of 1977. She expected to spend a year, but she ended up spending six.

There are three important research principles reflected in Barker's negotiations with the Moonies, each of which rests on the more fundamental principle that science must be objective, and objectivity requires intellectual freedom.

The first of these is *maximize your access.* Field researchers should not agree to limit their study to persons and places selected by others. Moreover, they should negotiate for recognition as a legitimate researcher with explicit permission to conduct interviews and to have access to all pertinent sites.

Second, *do not grant rights of censorship.* It is customary to offer to let those whom you study in the field read drafts of your reports—this often results in important corrections and may reveal important facts the researcher missed. But social scientists must make the final judgment about what to report.

Third, *secure independent funding.* Had Barker accepted financial support from the Moonies while she studied them (and she may well have been able to obtain such support from them), the objectivity of her results always would have been suspect and her freedom to observe and to report may well have been limited.

If it is impossible to meet these conditions, it is doubtful whether the study should go forward.

Neutralizing Observer Effects

It is widely believed that covert observers are much less likely than overt observers to change the behavior of those being observed. The idea is that, if people are not self-conscious they will act normally. There is less here than it would seem because people tend to respond to those around them and, even if they don't know they are being observed by a social scientist, they do know that *someone* is looking at them. Indeed, it now is recognized that most of what was observed by covert observers in a famous study of a small cult (Festinger, Riecken, and Schachter, 1956) was caused by the arrival and participation of new members, all of whom were psychologists from the University of Michigan (Bainbridge, 1996).

In any event, the disturbances caused by the presence of known observers can be neutralized, or at least identified. Here are some useful tactics.

First, stay long enough so that people become accustomed to having you around. In a sense, boredom is a fieldworker's best ally and often can cause people to forget that an observer is an outsider. Eileen Barker (1984:23) reported that, in time, many Moonies came to regard her as "a rather dull but harmless presence, to whom one did not have to pay much attention."

Second, keep a rather low profile in the beginning as you are learning to fit in and as others get used to you. However, just sitting in the corner and remaining silent may not be low profile, but may instead draw undue attention. Indeed, as Schatzman and Strauss (1973:59–60) pointed out, people being observed by a very passive observer usually will attempt to break through these limits because it is uncomfortable to be observed by a stranger.

Third, do not begin to be too inquisitive and too intrusive until you are familiar with the norms governing such interactions among group members and have established yourself as a trustworthy person.

Once again, let's listen to Eileen Barker (1984:20–21):

I found that the role I played as a participant observer went through three distinct stages during the course of the study. First there was a passive stage during which I did very little except to watch and listen (doing the washing-up in the kitchen was always a good place for this). Next there was an interactive stage during which I felt familiar enough with the Unification perspective to join in conversations without jarring; Moonies no longer felt they had to "translate" everything for me, and those Moonies who did not know me would sometimes take me to be a member. Finally, there was the active stage. Having learned the social language in the first stage and how to use it in the second, I began in the third stage to explore its range and scope. . . . I argued and asked all the awkward questions that I had been afraid to voice too loudly at an earlier stage lest I were not allowed to continue my study. I could no longer be told that I did not understand because, in one sense at least, I patently did under-stand quite a lot—and I was using Unification arguments in my questioning. . . .

Of course, even in the interactive stage it was known that I was not a Moonie. I never pretended that I was, or that I was likely to become one. I admit that I was sometimes evasive, and I certainly did not always say everything that was on my mind, but I cannot remember any occasion on which I consciously lied to a Moonie.

OBSERVATION

Regardless of the role a field researcher adopts, the brute fact remains that observation requires an immense amount of patience and concentration—it can be very hard work, intermixed with periods of intense boredom. Indeed, Barker (1984:21) reported that she often felt lonely, even though surrounded by Moonies, many of whom she liked.

Moreover, observation is difficult—even in very simple settings, no one can pay attention to everything. When a study begins with reasonably specific aims, it is easier to decide what to attempt to observe. But, even when a fieldworker begins a study with unspecified intentions, choices about what to observe must be made.

Most field research, at least during the early stages, is based on relatively unstructured observations. But, to the extent that the aims of the study have been carefully defined or as early observations focus the study, the observations can become more structured.

UNSTRUCTURED

Unstructured observations are informal, often impromptu, and usually are recorded in a narrative fashion.

When Lofland and Stark first began to observe the Moonies, the field notes they wrote down after each visit recounted everything they could remember about what had gone on, paying particular attention to facts about the personal histories of those present. Considerable space was given to describing how the Moonies organized and conducted their worship services, how they behaved when they ate together, and the like.

STRUCTURED

Structured observations are more focused and intentional and often are recorded on forms prepared for that purpose. For example, Lofland and Stark kept careful attendance records for Moonie services which led to their discovery of regular cycles in the number of newcomers present (Lofland, 1966). Or, having read claims in the British press that Moonies suffered from sleep deprivation and psychological pressure to such an extent that female members ceased menstruating, Eileen Barker kept track of the rate at which supplies of sanitary napkins were being replaced, finding that all seemed to be normal.

In even more structured observational studies, special forms guide not only what is to be observed, but how it is to be recorded as well.

Interviewing in the Field

Authors of books on how to do field research frequently claim that their method is superior to all others because it relies on observation—on things actually seen, not things "reported." Thus, John Lofland (1976:10):

> . . . *if social science is the study of what people actually do—if it is indeed the study of social action and interaction—then an empirical science must directly observe what it proposes to study! To propose to study something is to commit oneself actually to look at it; to stand close to it and to scrutinize it; to go where one finds it and watch it. Direct, empirical observation is by definition the basis of social, or any, science.*

But, despite such exhortations, observation is *not* the primary source of data field researchers obtain. It is not primarily through observation that anyone learns what participants in some social activity knew, felt, believed, remembered, hoped, understood, ignored, feared, distorted, or wanted. We do not read minds. What observational sociologist "sees" life histories? Indeed, large portions of what most field researchers report obviously took place when the observers were absent. That is, much of the most vital data produced by field researchers is not gathered by a sociologist who *watches* but, rather, by one who *listens* and who *asks questions*. *Interviewing* is the primary tool of field research. Let us, therefore, consider important aspects of interviewing in the field.

Informants and Respondents

People included in a survey are referred to as *respondents* because they respond to the interviewer or the questionnaire. For the most part, respondents are asked about themselves—about their background, about what they think, and about various kinds of behavior such as watching TV or reading the newspaper. Field research also makes considerable use of respondents—if usually less systematically. But a lot of field research is based on interviews with *informants*.

Informants are persons assumed to be well informed on matters of interest to a field researcher and who, therefore, are asked to provide information about a group as a whole or about particular members or subgroups.

Thus, while conducting his famous study of a youth gang in Boston, **William Foot Whyte** (1943) developed a special relationship with "Doc," the leader, and relied upon him for a great deal of information about other members and about events Whyte had not observed. In his equally famous study of African-American men who hung out on a particular corner in Washington, DC, **Elliot Liebow** (1967) relied on a man named Tally to help him understand what was going on. Eileen Barker (1984) acknowledged that much vital information came from members she referred to as her "moles." But it probably is anthropologists who have been most dependent on informants during their field research—their published studies typically acknowledge their debts to various informants.

Informants are a very valuable research resource. They can save researchers a great deal of time. For example, it might be very difficult and perhaps even impossible to discover the meaning and purpose of many customs and rituals. But an informant can quickly and easily clarify everything. Informants also may be able to provide information that an outsider is not permitted to observe—details of initiation rites, for example. Finally, field researchers can try out their various conclusions and interpretations on informants.

But informants also can be a source of misinformation and bias. Often, the persons most easily recruited as informants are marginal to the group and harbor various resentments. It is advisable to have a number of informants and to cross-check everything they tell you. Perhaps the major shortcoming of the use of informants in field research is to rely on them for information they cannot possess. Far too often, our information on proportional facts come from informants who could not possibly know the answers to questions such as: When is the average infant weaned? How many females engage in premarital intercourse? How many husbands beat their wives? Unless the answer is completely obvious as in a society where all husbands beat their wives, or all females engage in premarital sex, the informant can't know the answer *unless* he or she has done a survey or a census.

SYSTEMATIC OR IMPROMPTU INTERVIEWING

Most field research is based on unsystematic, impromptu interviewing. Thus, although the field researcher has certain interests, most of his or her interaction with respondents is "natural," in the sense that it is unplanned and spontaneous. This probably is the most appropriate style of interaction at the beginning of the research. It puts people at ease and lets them become used to having the observer around. It also maximizes the opportunity for unanticipated discoveries.

However, as the researchers begin to shape some specific conclusions it becomes appropriate to seek comparable data from everyone (or a sample). For example, Eileen Barker noticed that Moonies seemed to be unusually idealistic (in terms of their dedication to improving the world) and that this was not the result of their conversion, but because they had been brought up by unusually idealistic parents. This "fact" suggested that converts were attracted by the opportunity to fulfill these ideals, but before she could justify such a claim she needed first to discover whether or not the "fact" was factual.

To establish such proportional facts, the least field researchers can do is to raise this topic systematically with others. Better yet, they can do as Barker did: She conducted a field survey of Moonie members, asking a number of questions about their parents and their relationships with their parents.

FIELD NOTES

It does no good to go into the field to see things as they really are unless you carefully record what you see and hear, as well as your interpretations of what's going on. If you don't write it down, it's soon gone forever. Good field researchers write down or dictate **field notes** *regularly* and as *soon* as possible.

It often happens that researchers take far more notes at the start of the study when they are still trying to get oriented and to formulate more specific research questions. Moreover, much that goes on usually is quite repetitive and, once fully noted, need not be noted repeatedly.

In any event, the primary problems concerning field notes are not so much about taking them as about giving them systematic organization so that they can be efficiently retrieved and so that systematic conclusions can be drawn from them.

ORGANIZING FIELD NOTES

In the days before personal computers, field notes posed an immense problem of management. Most field researchers typed up their field notes after every period of observation—or had their dictated notes typed by someone else. Then, in the days before photocopy machines, field researchers had to construct elaborate indexes to their field notes. Thus, when they wished to examine all episodes of conflict within a group, for example, they had to leaf through their set of field notes, locating each page indicated in the index under the heading "intragroup conflicts." Once photocopying was available, multiple copies could be made and these could be cut up and filed under a number of topics. Thus, by opening the file folder labeled "intragroup conflicts" the researcher could save the time of looking up each incident and could proceed to organize these incidents on the basis of further distinctions, if appropriate.

Once PCs became popular, things got much more efficient. Now it is possible to append extremely elaborate coding schemes to each paragraph or textual block and to then sort and retrieve these in many different combinations. This can be done using any of the sophisticated word processing programs. There also are several programs available that were written specifically for field research notes.

ANALYZING FIELD NOTES

The ability to sort field notes is the basis of analyzing them. First, by sorting material into categories, it is possible to examine and compare all incidents of a particular process, event, or phenomenon. Second, it is possible to search any such set of materials for common patterns across incidents or to discover the need for additional categories. Third, sorting makes it possible to count the frequency of various kinds of events. How often did Moonies express doubts? Finally, it is possible to further subdivide categories. In how many of these instances were doubts privately expressed only to Eileen Barker? How many were expressed in front of other members?

Sorting also makes it possible to construct cross-tabulations and thus to seek correlations. Were people who expressed doubts more likely to quit than were others not heard to express doubts? Of the doubters, how many were leaders as compared with those who did not express doubts? How many had other family members in the group?

Often, it is useful to create many such cross-tabulations in order to search for causal relationships. For example, Lofland and Stark could identify strong, interpersonal attachments between each convert and a Moonie, existing *prior* to their conversion while noting that many people who attended a few meetings, but who did not convert, lacked such an attachment. This was the basis of their claim that attachments were a necessary, but not sufficient, cause of conversion (many who did not join also had attachments to Moonies).

STUDYING CHANGE

Field research often produces studies of change. Some of these are based on extended observations during which significant changes were observed. Others occur when a group is restudied, either by the same observer at a later date or by a new observer.

Field research usually is not a rapid style of research. The typical researcher probably spends at least 2 years in the field. Some spend more—Eileen Barker spent 6 years with the Moonies before writing her book, and Napoleon Chagnon has, by now, devoted 20 years to the Yanomamö.

By staying a long time, field researchers sometimes can observe changes. Thus, during Barker's stint with the Moonies she saw the process by which the group introduced a new type of membership and new organizations to serve them—the "home church" designed to let Moonies come together for worship, but without devoting full-time to serving the movement. This change reduced the rate of defections. But it also allowed some members to cut back on their participation who would probably not otherwise have done so.

Of course, in order for field researchers to take their time, their objects of study must have some degree of permanency. A study of a junior soccer team could not go on for years, nor could a study of most rock bands. Fortunately, those groups probably of greatest interest to social science tend to endure. Their endurance makes it possible for studies to cover a substantial period of time because these groups can be studied by a series of researchers.

When Eileen Barker went into the field to study Moonies, she already had read two substantial books to inform her about how things "used to be." The first was John Lofland's (1966) book (Stark dropped out of the study once the work on conversion was completed). The second was a book by David G. Bromley and Anson D. Shupe, Jr. (1979), based on observations of the Moonies ten years after Lofland's. While both of these books were based on fieldwork done in the United States, they offered careful documentation of many Moonie doctrines and practices that Barker found to have been modified in the interim. Consequently, anyone undertaking a field study of the Moonies today would have a careful record covering the previous 35 years.

ETHICAL CONCERNS

As noted, serious ethical issues surround field research. The first involves the right of *privacy*. It is nearly impossible to do field research, even as a known observer, without learning many things about people that they regard as private. An ethical researcher will take great care not to violate the privacy of those being studied. However, often it is precisely private matters that are of the greatest significance to the research. That some members of a work group hate the boss, sleep on the job, or cheat on their spouses may be critical data.

To solve this dilemma, field researchers often resort to the use of fictitious names, not only of people, but also of groups and places. Thus the research literature is abundant in made-up place names such as Middletown, Yankee City, Westville, and Lake City. It also abounds in made-up group names: California cult, the corner boys, and the Blue Gang. Lofland and Stark initially shielded the Moonies behind the name Divine Precepts and used a pseudonym for each group member. And, true to her word, Eileen Barker did not reveal information about rank-and-file members that would have embarrassed them, not only to outsiders but, even more importantly, to other group members.

Beyond privacy lies the issue of *deception*. How far can a field observer go in concealing his or her true identity or motives? The temptation to mislead is very great in field research, because by doing so one can observe groups that would never permit a known observer among them, or one can gain confidences that would not otherwise be given. These are matters on which people of good character can disagree. But human subjects committees tend to be extremely restrictive in the use of deception by social researchers.

A third ethical problem many field researchers confront is that, often unwittingly, they *contribute to* and even *take an active part in the activities they observe*. For example, during a field study of a big-city police department, Jerome Skolnick (1966) not only often was mistaken for a detective, but the police came to rely on him as an assistant, asking him to drive a vehicle during raids, for example. John Lofland was expected to help the Moonies copy edit the English translation of their scripture. And because Eileen Barker has such a regal persona (before becoming a social scientist she was on the British stage and her husband was for many years the BBC radio and television voice of the London Symphony), by her mere presence she lent prestige to Moonie gatherings.

Such involvement seems innocent enough, but it becomes much less so when those being studied engage in illegal activities. To have been present during a drug sale or when a youth gang plans a crime is to be *an accessory* in the eyes of the law. How does one study such activities and not be an accessory? By relying entirely on interviews and being careful never to be an observer would reduce the legal liabilities. But even to know of guilt and not to report it is illegal. Unlike lawyers, psychiatrists, and the clergy, social scientists have no legally protected confidentiality. They can be called into court and required to testify.

A somewhat less direct ethical concern involves maintaining one's objectivity and intellectual independence despite spending long periods in close, often intimate, contact with some set of people. It is normal for people to form strong bonds of friendship and even affection as a result of extended interaction. Thus, many social scientists have become very outspoken advocates of groups they studied. Even more dramatic is conversion—sometimes, field observers become committed participants in the groups they set out to observe. Several social scientists joined the cults they were studying, and it is widely rumored among anthropologists that several who went off to study a tribe, stayed.

However, a field researcher also risks losing his or her objectivity in the negative direction as well. Often this occurs because the stresses and strains of doing field research color the observer's views. As **William Shaffir**, **Robert Stebbins**, and **Allan Turowetz** (1980:3) explained:

> *Fieldwork must certainly rank with the more disagreeable activities that humanity has fashioned for itself. It is usually inconvenient, to say the least, sometimes physically uncomfortable, frequently embarrassing, and, to a degree, always tense.*

Indeed, it caused a considerable scandal when, long after his death the diaries that Bronislaw Malinowski kept during his years among the Trobriand Islanders were published. The diaries covered the years when, among other things, he developed the theory of magic discussed in earlier chapters. The diaries revealed that Malinowski had a considerable dislike for most of the islanders— "They lied, concealed and irritated me. I am always in a world of lies here" (in Lofland and Lofland, 1984:34). Did his negative feelings shape his analysis? Since he is our only source of information about what he saw and heard, there is no way to tell.

CONCLUSION

Perhaps the best summary of all the issues and problems facing field researchers is "Welcome to the real world."

Field research allows us the opportunity to see significant social phenomena as they are, not as they might appear from the library, from the laboratory, or from questionnaire responses. On the other hand, because fieldwork is done out in the real world, field researchers are exposed to many of the normal stresses and strains that go with the territory. Moreover, because field researchers so often study unusual groups, those they are observing may try to force them to do things they don't want to do, can't do, or which are illegal—indeed, field researchers have been seduced, threatened, humiliated, raped, beaten up, and possibly murdered (Lee, 1995).

Being up close is not only dangerous, it also can offer a very limited scope of vision. We may be in the wrong place when a crucial event occurs. Informants often lie. People will offer honest and plausible accounts and interpretations that are, nevertheless, completely wrong. We may see things the way we wish them to be. Finally, with no one looking over our shoulders and with glittering opportunities beckoning, we may report things that aren't so.

But those social scientists who embark on field research with significant questions in mind, who have successfully negotiated proper ground rules concerning their access, who use not only informal observational methods, but any methods that are appropriate for the questions that arise, who endure the rigors and boredom of the task long enough to get good data, and who then analyze and report their results carefully, fully, and honestly are able to make important contributions to our understanding of how the world works. And that's the whole point—isn't it?

REVIEW GLOSSARY

- **Unstructured** observations are informal, often impromptu, and usually are recorded in a narrative fashion.

- **Structured** observations are more focused and intentional and often are recorded on forms prepared for that purpose.

- **Informants** are persons assumed to be well informed on matters of interest to a field researcher and who, therefore, are asked to provide information about a group as a whole or about particular members or subgroups.

- **Field notes** are written accounts of what a field researcher sees and hears. They should be made as soon as possible after the observations and should be as complete as possible.

Experimental Research

Recall from Chapter 6 that experiments have two essential characteristics, and studies lacking either are not experiments. First, the independent variable or variables are *manipulated*. Second, subjects are *randomly* assigned to be exposed to different levels of the independent variable. This means that groups should differ only in terms of their exposure to the independent variable, all other differences being randomly distributed. The importance of these two aspects of experiments is that they control time order and eliminate potential sources of spuriousness.

Here we pursue these matters at far greater length. As we shall see in this chapter, there are many research designs that often are called experiments but which either fail to manipulate the independent variable or do not use random assignment of subjects—and some fail to do either. We will discover why such studies leave much to be desired.

Before taking up these matters, however, it is important to identify some basic applications of the experimental method based on the typical settings in which experiments take place.

EXPERIMENTAL SETTINGS

The word *experiment* conjures up images of scientists in white lab coats working in carefully isolated and insulated laboratories. They might be injecting rats with a new medication or supervising college students assigned to learn various lists of nonsense syllables.

It is true that some experiments are conducted in laboratories. It also is true that many of them—even experiments testing new drugs—take place in the field. In fact, some experiments are conducted during survey interviews. Let us examine an example of each experimental setting and assess its typical advantages and disadvantages.

LABORATORY EXPERIMENTS

Perhaps the majority of social psychological experiments take place in laboratories located in universities. Students usually are the subjects. In addition to being convenient, laboratories offer a maximum control over all relevant factors. Labs can be soundproof. They can be closed off to prevent interruptions. Subjects can be enclosed so that their ability to observe is controlled. To illustrate, let's examine a very important and well-designed experiment.

In a classic study known as the original bystander intervention experiment, **John Darley** and **Bibb Latanè** (1968) set out to explore the conditions under which individuals do or do not take action in response to a perceived emergency. They believed that, the larger the group observing the emergency, the more likely any given individual was to rely on someone else to act in response to the emergency—a tendency they referred to as the "diffusion of responsibility."

Darley and Latanè began by recruiting first-semester college students to participate in a study. Upon arrival each student was shown to a booth equipped with a microphone. Each was to be a participant in a panel discussion about the problems of adjusting to college. Students were told that, to maintain anonymity, the discussions would be done over an intercom system and a red light would come on in their booths whenever it was their turn to talk.

In this experiment, the independent variable was perceived group size—variations in the number of others believed to be present. Some students were told they would be one of six persons on the panel. Others were told they would be part of a three-person group. Others were told that they would have only one discussion partner. They also were told that the social scientists would not be present during the discussion and would not be listening in.

As each discussion began, during the first go-around, one of the members in each group mentioned being anxious about being in college because he sometimes had epileptic seizures. The next time this student came onto the intercom system, he suddenly began to stammer and breathe hard. Then he gasped out, "Help," followed by the sound of a body falling to the floor. No one else could be heard because the victim's mike was still on. Who would leave their booths and seek help?

All students who believed they were the only other person aware of the emergency left their booths and went for help. But only 80 percent of those who thought one other person knew (who thought they were in a three-person group) went for help. Only 60 percent of those who thought they were in a six-person group left their booths in search of help. We know these statistics because, of course, Darley and Latané were peeking and recorded who actually went in search of help. By the way, each student actually was alone. All of the other voices, including the episode involving the seizure, were tapes.

The power of the experimental method lies in the fact that in Darley and Latané's laboratory *only* the independent variable varied. *Everything else* was held constant—made the same for all subjects. Each subject was given exactly the same instructions (the experimenter had memorized them)—except for the number of others said to be taking part. Every subject heard the same tapes (except for the number of voices they heard). Everyone heard the same tape of the seizure. Everyone was in the *same* isolation both. Finally, because the independent variable was manipulated randomly, the groups thinking it was a two-person, three-person, or six-person panel should have been alike in terms of their personal characteristics.

As a result, we can place very great confidence in the conclusion that variations in group size *caused* the variation in the percent who went for help. But, if the strength of the laboratory as the experimental setting is the immense amount of control the experimenters have, therein also lies its major weakness. Many people cannot be brought into a laboratory or can't be kept there sufficiently long, and many important matters simply are ill suited to this setting.

FIELD EXPERIMENTS

Few people today can recall a polio epidemic. But, until the middle 1950s, public swimming pools and beaches often were closed (sometimes even the movie theaters, too) because of outbreaks of polio. This dread disease often was called *infantile paralysis* because it primarily struck children. While only about 1 in 20 victims died, many were left paralyzed (some unable to breathe without respirators) and many more suffered from withered and deformed limbs.

President Franklin D. Roosevelt was a polio victim who had lost the use of his legs (he was a rare, adult victim of the virus). Soon after his election, he decided to create a private, charitable organization to provide treatment for polio victims and to seek a cure. The organization became known as the National Foundation for the March of Dimes because the fund-raising campaign stressed how much could be accomplished if each person gave only a dime. Soon, the March of Dimes was funding research in dozens of laboratories, each racing to be the first to develop a vaccine (Smith, 1990).

By the early 1950s, a polio vaccine was in sight. Dr. Albert Sabin at the University of Cincinnati was thought to be within five years of perfecting a vaccine using live viruses. Then Dr. Jonas Salk at the University of Pittsburgh announced he had produced a vaccine using dead viruses. Salk acknowledged that his vaccine had not reached the point of 100 percent effectiveness, but that it probably would reduce the incidence of polio by about 75 percent. Many polio researchers advised the March of Dimes officials to wait for the Sabin vaccine, because it would be 100 percent effective. But the officials refused to wait on the grounds that lives and bodies were at stake. Even a partially effective vaccine would be far better than nothing (Smith, 1990). But did the Salk vaccine provide even partial protection? How could this be determined?

Suppose the March of Dimes had simply begun a nationwide program to give the vaccine to everyone. Would it not soon have been evident that the incidence of polio was going down? Perhaps, but polio was subject to very substantial fluctuations over time having nothing to do with treatments or vaccines. Sharp declines lasting for several years were common. What about seeing how many people who had been vaccinated came down with polio? But Salk's vaccine was not expected to be 100 percent effective. So, if some people who had been vaccinated got polio, was it because the vaccine was ineffective or because it wasn't completely effective?

The solution was to do an experiment in which one group made up of persons randomly selected to get the vaccine was compared with another group of persons who did not get the vaccine.

An **experimental group** consists of those persons randomly assigned to be exposed to the experimental stimulus—to the independent variable.

A **control group** consists of those persons randomly assigned to not be exposed to the experimental stimulus. The purpose of the control group is to provide a baseline against which to assess the effect of the experimental stimulus.

The techniques for *randomly* assigning subjects in experiments are the same as those for random selection in sampling, discussed at length in Chapter 4. Persons can be assigned to groups by randomly generating numbers, drawing disks from a drum, or flipping a coin—so long as the procedure used is truly random.

When the independent variable takes several values, rather than being merely present or absent as in the case of a vaccine, technically there is no control group in the sense of one group that is not exposed to the experimental stimulus. Instead, comparisons are made of groups differing in the level of the experimental stimulus (or independent variable) to which they were exposed—group size, in the Darley and Latanè experiment. When this is the case, rather than refer to an experimental and a control group, social scientists refer to *treatment* groups and identify each on the basis of its level of treatment.

It would not have been possible to do an experimental test of the Salk vaccine in a laboratory because a large number of subjects were needed and they would have to be studied for a long time. The incidence of polio per year was about 50 cases per 100,000 population. Thus, if 40,000 people were included in the experimental group and given the vaccine, while another 40,000 were included in the control group and given a harmless substitute for the vaccine, then, if the vaccine had *no effect*, there would be only about 20 cases of polio in each group. But, if the vaccine were 50 percent effective, then we would expect about 10 cases in the experimental group and about 20 cases in the control group. If it were 100 percent effective, there would be no cases in the experimental group and about 20 in the control group. These numbers would be far too small to give a trustworthy answer to whether the Salk vaccine was effective. Because of the urgent concern to prevent polio, the largest field experiment in history was designed and conducted—and without any government funding (Smith, 1990).

In the fall of 1954, the March of Dimes made arrangements to test the vaccine on nearly 2 million schoolchildren from around the nation (Brownlee, 1955; Meier, 1972). Local physicians were recruited to give each child a shot. On a random basis, half of the children received the real vaccine and the other half a harmless salt solution, but the physicians giving the shots were unable to tell when they were injecting the real thing or the *placebo*.

As used in experiments, a **placebo** is a substitute for the independent variable and is used to permit variation in the independent variable without permitting variation in other factors linked to the independent variable. For example, in an experiment to test a drug, one treatment group gets the drug while the other gets a harmless substitute, but *both* groups get injected.

In this instance, the placebo was used to make certain that the effect was due to the drug, not just to being injected with something by a doctor. This is important because results often show that simply being treated can have an effect on the dependent variable, an effect that has nothing to do with the actual independent variable—this effect often is referred to as a *placebo effect*. Thus, for example, a number of treatments seem to help various physical conditions, but the use of a placebo reveals that the improvement is nothing beyond the placebo effect—that doing *anything* for a patient will cause a significant number of them to get better. By using placebos, researchers can determine whether the experimental group displays an effect greater than that produced by the placebos, thus eliminating the placebo effect from their calculations.

The reason that the physicians were prevented from distinguishing the Salk vaccine from the placebo was to make sure they acted the same with each child. Perhaps doctors would have felt some guilt or resentment about giving a child a "worthless" shot rather than the real vaccine had they known when they were doing so. But they couldn't influence the results with their reactions because they didn't know. An experiment with this type of safeguard is called a *double-blind experiment*.

A **double-blind experiment** is one in which neither the subjects nor the experimenters know whether or when the independent variable varies.

Following the administration of the shots, the researchers kept close track of the subjects to see who contracted polio during the next year. Every reported case of polio was reviewed by a panel of experts to make certain of the diagnosis. The experts also separated the cases into paralytic and nonparalytic forms of polio, because it was not certain that they were precisely the same disease. The results showed that there was a striking and highly significant difference when only the paralytic cases were compared. Thirty-three of the vaccinated children developed paralytic polio as compared with 115 of those who had received only the placebo. Moreover, of those who contracted paralytic polio, four of those who had received the placebo died while none of those who had been vaccinated died. However, those who had been vaccinated were equally likely to contract nonparalytic polio as were those who had received the placebo (Brownlee, 1955; Meier, 1972).

The Salk vaccine worked (if not perfectly) for paralytic polio. Soon after, the Sabin vaccine became available, and a nationwide campaign succeeded in immunizing nearly everyone.

Although drug testing often involves field experiments, many other topics also have been the object of field experiments. Field experiments are an especially effective way of evaluating all sorts of public programs. For example, in Minneapolis, police officers responding to a call concerning domestic violence were first instructed to determine whether it was a case in which they were *required* to make an arrest—had there been serious bodily harm or use of weapons? If so, the suspect was arrested. If not, the suspect became an experimental subject. Officers were randomly instructed to arrest the offender, to force him to leave, or to attempt to mediate the dispute. Thus, the independent variable took three values. Each case was then closely monitored (victims were frequently interviewed) for a year to see whether new violations occurred (the dependent variable). The results showed that men who were arrested were far less likely than the others to reoffend (Sherman and Berk, 1984).

Many field experiments have been conducted to evaluate various educational methods and teaching techniques. Delinquency prevention programs also have been experimentally evaluated as have a variety of social welfare programs.

EXPERIMENTS IN SURVEYS

Experiments are incorporated into surveys for several reasons. The most common reason involves the need to assess survey techniques as such. For example, techniques for minimizing refusal rates often are tested experimentally. The sample is randomly separated into two or more subsamples each of which is then approached with a different method. For example, a mail questionnaire study discovered that the response rate was significantly higher when a dollar bill was included with the questionnaire (Cook and Schoeps, 1985). Sometimes the experimental approach is used to assess different question wordings—you will examine examples in the exercises.

A second reason for the use of experiments in surveys involves the desire to see if the independent variable causes the dependent variable and to generalize that effect (in terms of size as well as direction) to a significant population. **Guillermina Jasso** and **Peter Rossi** (1977) wanted not only to see if various combinations of occupations and incomes caused people to rate some as very over- or underpaid, but also to generalize their findings to the general population. Thus, each respondent in a survey was presented with a randomly generated set of descriptions of persons having a particular job and a particular income. For each description, respondents were asked whether the salary was far too high, too high, about right, too low, or far too low. The amount of income associated with any specific job was varied randomly from a very large amount down to a very small amount. The results showed that people had strong and very consistent standards about which jobs should receive high, medium, or low salaries. Because the findings were based on a survey of the adult population rather than of student volunteers, the results have greater descriptive application.

RELIABILITY

Reliability concerns consistent measurement. A primary method for assessing reliability in experiments is replication. Because it is inexpensive to do so, any laboratory experiment that produces results regarded as important most certainly will be replicated—often many times. Failure by replicators to obtain reasonably similar findings is taken as proof that the original findings were unreliable—perhaps for purely random reasons but more often because of bias (accidental or intentional). There are two primary sources of bias that can produce faulty results. If these biases occurred in the initial experiment but then were eliminated in replication studies, the results would not be reconfirmed. Sometimes, however, the replicators fail to see the biases operating in the original experiment and repeat these errors, in which case the results often are confirmed.

SUBJECT BIAS (DEMAND CHARACTERISTICS)

The placebo effect is an example of how bias may influence subjects. If the people in the experimental group know they are being treated for some physical symptoms, but those in the control group know they are not being treated, that knowledge alone often is sufficient to make it appear that the treatment is effective. Thus, it is necessary for both groups to believe they are receiving treatment in order to eliminate this source of bias.

A famous study of worker productivity found that all changes in the environment produced increases in productivity. For example, when the lights got brighter, people got more work done. As the lights were dimmed, they continued to do even more work (Roethlisberger and Dickson, 1939; Homans, 1950). The experimenters had intended to use variations in such factors as lighting as independent variables. It turned out, however, that, no matter what they tried, any attention from social scientists was what the workers paid attention to and that was the actual independent variable. The workers tried as hard as they could to meet the expectations of the social scientists. This phenomenon is often called a *demand characteristic*.

In experiments, a **demand characteristic** is an aspect of the design that tips off subjects as to what they are expected to do. Subjects tend to be cooperative and, therefore, to respond to demand characteristics.

The primary solution to this source of bias is to make sure the subjects are unaware of what the experimenters expect or hope will happen. Sometimes, that means that the instructions they are given should draw subjects' attention away from the independent and dependent variables. Often, however, especially in field experiments, this is not possible and great care must be taken to minimize demand characteristics.

EXPERIMENTER BIAS

Remember that in the polio experiment the physicians were kept unaware of which shots were real and which were placebos. This was done to prevent experimenter bias. When the people conducting an experiment know who is about to receive what level of the independent variable, they often unwittingly alter their behavior in ways that can influence the outcome. Thus, they may act friendlier or give the instructions more carefully and slowly to people in one condition as opposed to those in another. This problem becomes especially acute in field experiments when the relationship between subjects and experimenters is a long-term one. Thus, for example, teachers using a new method may be more enthusiastic and work harder than teachers using the old method. Therefore, the experimental results show that the new method is far superior. But later, after the method has been adopted for all classes, everyone is surprised to discover that the expected improvements have vanished—it was the quality of teaching, not the method, that was the actual cause of the experimental differences.

MEASUREMENT VALIDITY

As will be clear, the term *validity* has far wider meaning among experimentalists. This is because of the power of the method for demonstrating causation—if things are done right (in a valid way), then the results must be valid. Consequently, in this section, the discussion of validity will focus narrowly on issues of measurement validity.

ASSESSING THE MANIPULATION

Researchers once informed teachers in various elementary schools that test results showed that some of the students who would be in their classes the next fall were due for a "learning spurt" (Rosenthal and Jacobson, 1968). That is, these students would suddenly begin to do much better in school than they had in the past. The students so identified were selected randomly. The hypothesis was that teachers would alter their behavior toward these students in ways that would cause them to actually improve on their previous school performances. But, after the school year was over, the students identified as due for a learning spurt had not outperformed their classmates.[1]

It turned out that these results had no implications for the hypothesis one way or the other, because the teachers were not able to identify which students had been predicted to do better. The manipulation of the independent variable had been based on a brief memo placed in each teacher's box near the end of the school year. Those who glanced at it had not yet encountered the students. Most thought that, if the memo was really important, as compared with all the other memos they were receiving, they would hear more about it. Many couldn't even recall the memo in the fall. Consequently, the independent variable in this study *did not vary*. Teachers could hardly have responded to the knowledge that some kids were slated for a spurt, because they never knew.

Thus, it is important to make sure that the independent variable really varies—that the manipulation is effective. Recall the experiment designed to see if gender influenced political choices (Chapter 6). Perhaps the lack of gender effects was the result of people not having noticed gender differences between the two candidates.

Good experiments attempt to check on the manipulation. In the case of the learning-spurt experiment, teachers should have been notified again in the fall when they had actual students with whom to connect the information. Then, they could have been asked for an initial impression of each of their students and to indicate which of them was predicted to have a learning spurt. Or, in the case of the gender and voting study, after completing the experiment, subjects could have

---

[1] Initially, the researchers reported that these students had, in fact, outperformed others as hypothesized. Subsequent reanalysis of their published findings showed that they were mistaken and revealed their failure to successfully manipulate the independent variable.

been asked about the gender of the candidate for whom they voted and about the gender of the other candidate. This questioning could not influence their responses, since these already had been registered, but it would have demonstrated the effectiveness of the manipulation.

REALISM

A major criticism of experiments is that they are artificial and that, especially in the laboratory setting, the independent variables are not sufficiently real. This criticism is sometimes warranted. Ironically, the more that experimenters attempt to be realistic, often the more unconvincing are their claims to validity. To see why, let's examine an experiment that received a great deal of coverage in the news media and which is widely cited as an example of an experiment that overcame the "artificiality" of the lab setting.

Stanley Milgram (1974) wanted to study the social processes by which ordinary humans can be induced to do extraordinarily inhuman things to others. For example, why did German prison camp guards participate in mass executions without any protest? Milgram was convinced that the key to understanding such behavior lay in the dynamics of obedience to authority. He designed a series of experiments to explore this phenomenon.

In one of his experiments, each student subject was told that he or she would be participating in an experiment designed to test the effects of punishments on learning. Each subject was given the role of the teacher and was instructed to read a series of word pairs to a learner (who was introduced as another student subject). After reading the word pairs twice, the teacher would go back through the list, reading only the first word of each pair, testing the learner's ability to correctly recall the second word of each pair.

Each time the learner failed to supply the correct response, the teacher was instructed to administer an electric shock to the learner and then repeat the correct response. The learner was then strapped into an electric chair. The teacher stood in front of a power supply board on which there were 30 switches, each labeled according to voltage—from 15 volts to 450. In addition, the switches were grouped into seven sets of four and a set of two. Each group was identified by large labels ranging from "slight shock," "moderate shock," "very strong shock," "intense shock," and "extremely intense shock," "danger: severe shock," to "XXX." The teacher was instructed that, if the learner missed a pair, he or she should administer the lowest voltage switch. For each additional incorrect answer, the next higher voltage switch should be used.

An experimenter in a white lab coat stood next to each teacher to help out if needed. Then the learning session began. The pairs were quite obvious combinations such as "dog/cat." After going through the list twice, it was time to test. Surprisingly, the learner missed on the very first try. The teacher administered a shock by flipping the 15-volt switch on and off. The learner responded with an obvious flinch of discomfort. But the test continued, and the learner continued to make mistakes and to receive progressively higher shocks.

In fact, the learner was not actually a subject but another experimenter who always missed an average of two out of every three times. Moreover, the power supply board was phony—no power ran through it and no shocks were actually administered to the learner. However, as the misses began to pile up and the voltage was increased, the learner began to indicate pain and distress. At 75 volts, grunts and groans began. At 180 volts, the learner cried out, "I can't take any more." At 315 volts, the learner demanded that the experiment stop. Any time the teacher hesitated to give a shock, the experimenter in the lab coat, calmly and firmly, instructed the teacher to continue. At 315 volts, many teachers did not want to go on, but the experimenter told them, "You have no choice, you must go on." If the teachers refused to continue, the experiment was stopped. But it continued as long as the teacher would flip the next switch—all the way through 450 volts (also marked XXX).

Milgram conducted this basic form of the experiment many times, each time using a different independent variable. The most significant of these was *proximity*. Thus, some subjects were randomly assigned to a lab situation in which the learner was in another room so that he or she could be heard but not seen. In another situation, the learner was in the same room as the teacher but at a substantial distance. In a third situation, the learner was so close that the teacher could reach out and touch her or him. Variations in proximity did produce variations in obedience. Teachers were significantly less likely to give high-voltage shocks to a learner within reach, somewhat more likely when the learner was in the room but at a distance, and most willing to shock people they could only hear.

Milgram's results caused a media furor: A Yale professor had found that a majority of his subjects were willing to deliver high-voltage shocks to helpless victims, who were pleading with them to stop! Headlines suggested that Nazi prison guards lurked within most Americans. Milgram was praised for not restricting his work to insignificant, artificial laboratory studies and for dealing with real problems in real ways.

But look more closely and it will be clear that the reality of this experiment is very superficial and unconvincing. This was not a field experiment. These were not Nazi social psychologists trying to screen soldiers as to their suitability for prison camp duty. Every subject knew that he or she was in a laboratory in the psychology building of Yale University. The man in the white coat was not an S.S. officer with immense power over them but a Yale professor (or a graduate student). How many subjects would believe that they were being asked to murder or seriously injure someone as part of an experiment done in this setting by such people? In fact, those gullible enough to worry about this possibility were precisely those who refused to go on.

In other versions of the experiment, Milgram varied how much people could hear and then how much they could see. But it hardly took such an elaborate setting to demonstrate that people prefer not to see, hear, or be close to those they punish. That's really all we can tell from this set of experiments, for these are the only conditions that varied.

Later in this chapter, we will discuss the reality or artificiality of experiments in the more general context of external validity. But, to anticipate that discussion, keep in mind that theories are not limited to predictions about a "real world" that somehow excludes the laboratory. The lab is not in outerspace somewhere. Many theories make definite predictions about what should and should not happen in the laboratory as well as elsewhere. Further, many experiments take place in the field—as when real teachers try a real teaching method with real students.

In any event, the important questions about the validity of experiments involve the validity of their basic designs.

ASSESSING EXPERIMENTAL AND QUASI-EXPERIMENTAL DESIGNS

Many things done by scientists are called experiments. Some of these things are real experiments, and some aren't. Let's sort them out, starting with the least adequate designs. For economy of presentation, each design will pursue the same basic research question. Some social scientists have produced a set of movies they think will reduce racial prejudice. Now they want to assess their effectiveness experimentally. The *independent* variable will be the movies. The *dependent* variable will be measured by a questionnaire designed to assess racial prejudice.

DESIGN 1

In this design, the social scientists persuade the principals of some grade schools to take part in the study and then request permission from each student's parents for the child to watch the movies. During the next six months, each child whose parents gave permission stays after school once each week to watch a 30-minute movie. At the end of the six-month period, each child is given the questionnaire and the average prejudice score turns out to be 25 (on a scale of 0 to 100). We can diagram this design this way:

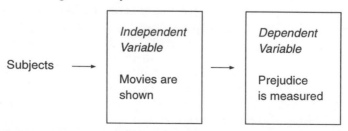

What can we learn from this study? Absolutely nothing! The independent variable did not vary. Every subject saw the movies. Would they have scored higher or lower or the same had they not seen the movies? No answers are possible because no comparisons are possible.

Design 2

Let's try a better design. Each student's prejudice is measured before as well as after seeing the movies. Now students' scores after seeing the movies can be compared with their scores before seeing the movies.

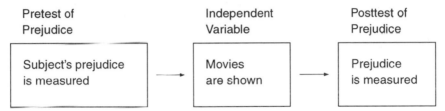

Now a comparison is possible, and it turns out that the average score was 50 before the movies were shown and dropped to 25 following the movies. So we can be sure that the movies do cause racial prejudice to decrease. Can't we?

History.   Certainly not! For one thing, six months have passed. Other factors, things not controlled in this design, could have been the true cause. One of these factors is *history* (Campbell and Stanley, 1963).

Used this way, **history** refers to any event occurring simultaneously with the independent variable that might account for the observed changes.

It took six months for the students to see the movies. What if, during that same period, the first students of another race enrolled in the school? What if a very dramatic incident involving racial tolerance were given extensive coverage in the news media? What if students read stories depicting how kids of another race feel bad when they are the targets of prejudice? We simply don't know what else may have happened to influence the scores.

Maturation.   Another uncontrolled factor is *maturation* (Campbell and Stanley, 1963).

In experiments, **maturation** refers to the effects of subjects getting older (maturing) as time passes.

After six months, the students are older than they were when they took the pretest. Maybe kids naturally start to outgrow prejudice at a certain age.

Testing Effects.   A third uncontrolled factor involves *testing effects* (Campbell and Stanley, 1963).

Testing effects occur when the tests or measurements are themselves causative agents—whether by changing views or by influencing performance, as in the case of subjects becoming testwise.

Sometimes, simply by being measured on some trait, people are changed in terms of that trait. A sad example was reported by a survey interviewer who was asking questions designed to measure anti-Semitism. The respondent agreed with the first several negative statements about Jews. Then, when the interviewer read the statement "International banking is pretty much controlled by Jews," the respondent said, "I didn't know they did that, too!" The interview itself increased this man's prejudice. But it is equally likely that, for some people, their prejudice can be reduced by answering questions such as these—that simply by having been asked the question, people begin to think more about these matters and, perhaps over a period of time, come to different conclusions. Consequently, the pretest may have been the true cause of the decline in prejudice.

Instrument Effects.   In some cases, responses also may change, not because the subjects changed, but because the *measuring instruments changed* (Campbell and Stanley, 1963).

Instrument effects refer to differences produced by variations in the accuracy of multiple measurements.

If the pretest is not as stringent a standard of prejudice as the posttest, then scores will decline.

Because this design does not control for any of these factors, we can't know whether these results were caused by seeing the movies or whether they would have occurred even if the kids hadn't seen any of the movies.

Now let's look at a slight variation of this design. Often researchers decide to focus on people who score extremely high (or low) on a pretest. Thus, an experiment designed to improve math skills or reading level might select students who were most in need of help, those with the worst achievement scores, and include only them in the experiment. In the example we have been using, the researchers might decide to show the movies only to the most prejudiced kids.

Regression to the Mean.   Suppose these researchers select only the most prejudiced students. Then, using Design 2 they discover that the average prejudice score declines from 90 to 80 after students have seen the movies. When only extreme scorers are included in an experiment, the possibility exists that the findings reflect nothing more than *regression to the mean* (Campbell and Stanley, 1963).

Regression to the mean refers to the tendency of extreme scores to move (or regress) over time toward the group average (or mean).

This is a well-known statistical principle applying to phenomena having natural variation (as opposed to being fixed on some basis such as physiology or law). When cases are selected on the basis of scoring far from the mean (whether above or below), the best prediction is that, on a subsequent measurement, they will score closer to the mean. Consequently, when results are measured by comparing the same subjects at two points in time, a change is to be expected. The question thus posed is, "Was the change entirely the result of regression to the mean, or did the movies matter, too?"

DESIGN 3

Perhaps we can overcome these problems by having a control group. Recall that parental permission was required to take part in the experiment. So let's compare those who saw the movies with those who did not because they did not get permission or did not want to stay after school to watch movies.

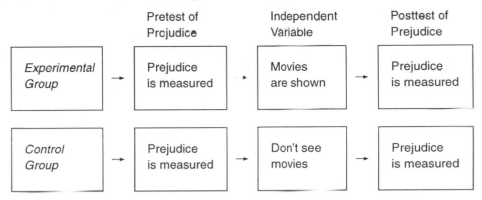

Here we needn't worry about many of the possible sources of bias that were uncontrolled in Design 2. Any history effect will show up in both the experimental group who saw the movies and the control group who did not. Differences between the two cannot, therefore, be caused by history. Neither can be caused by maturation or testing or instrument effects. In fact, if both the experimental and the control groups were made up of students with equally extreme scores, regression to the mean would be ruled out, too.

Nevertheless, we still can put no faith in the findings. The students who took part in the experimental group differ in obvious ways from those in the control group and may also differ from them in many other unknowable ways. That is, *selection bias* may be the source of the findings (Campbell and Stanley, 1963).

Selection Biases.   Whenever people are assigned to the experimental or control groups (or to treatment groups based on levels of the independent variable) in nonrandom ways, **selection biases** occur.

Perhaps parents who gave permission were more concerned about reducing their children's prejudice, and the movies prompted them to often talk about such matters with their children. Indeed, the kids who were willing to stay after school to watch antiprejudice movies may have been far more predisposed to change. Or they could have been more easily persuaded or more responsive to authority.

There is no sense in worrying about these problems. They are easily solved by randomization. In this example, we could simply ignore the kids who didn't get permission and randomly assign those with permission to the experimental or the control group. Should it seem inappropriate to simply exclude half of the subjects whose parents had given permission, subjects in the control group could be shown movies having nothing to do with reducing prejudice.

DESIGN 4

By randomly assigning subjects to the experimental and control groups, we are able to use tests of significance to tell us whether an effect is sufficiently large to be taken seriously—to not be the result of random fluctuations. Put another way, the purpose of random assignment is to make the experimental and control groups alike in every way except their exposure to the independent variable.

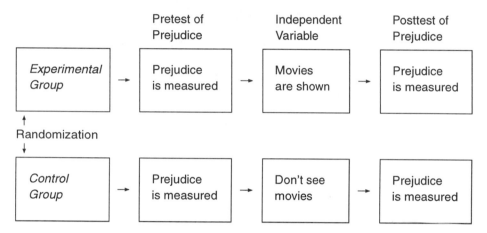

Only when randomization is the basis for selection is the study a true experiment. All three earlier designs can be identified as merely *quasi-experiments* (Campbell and Stanley, 1963).

Quasi-Experiments.   Many studies pose as experiments, but fail to meet the essential experimental standards of internal validity. These are called **quasi-experiments** (*quasi* means "having superficial resemblance to").

Internal Validity.   The standard of **internal validity** consists of eliminating all of the sources of bias. That means there must be at least two groups to compare (two levels of the independent variable including present and absent) and that assignment to groups must be random.

In most laboratory experiments as well as in many field experiments, there is no need for pretest measurement of the dependent variable. Random assignment should ensure that groups begin at the same average level. However, when an experiment has a considerable duration, pretests are needed to detect one additional possible source of bias: *subject "mortality"* (Campbell and Stanley, 1963).

Subject "Mortality."   The loss of subjects over the duration of an experiment is called **subject "mortality."**

Recall that our experiment on prejudice reduction requires students to attend movies over a period of six months. Not every student will be able to attend every time, and some students will drop out for a variety of reasons. There is no reason to suppose that either missing some movies or dropping out entirely is random. Therefore, differences between the experimental and the control group could be produced because of biases in who dropped out. For example, it is reasonable to suppose that the most prejudiced students in the experimental group would dislike the movies or find them disturbing and drop out. Pretest data make it possible to check for the existence of mortality effects—are the dropouts different from the rest in terms of their initial scores on the dependent variable?

The Fallacy of Unmanipulated "Causes"

Suppose that, after conducting the prejudice-reduction experiment, the researchers suspected that the effect had been greater among girls than among boys. So they separated the experimental and control groups into males and females and compared them. Suppose they found that the decline in prejudice scores was, in fact, greater among girls—both within the experimental group and by comparison with the control group. The researchers announce that the movies were more effective with females.

Such claims appear all the time as experimenters dredge through their data looking for additional causes. But they are *not* experimental results. Instead, they represent the *fallacy of unmanipulated causes*.

The **fallacy of unmanipulated causes** involves mistaking the effects of individual *characteristics of subjects* on the dependent variable for experimental effects.

An experimental effect has been observed if, and *only if*, the independent variable was *manipulated* on a *random* basis. Characteristics of subjects cannot be manipulated at all, let alone randomly. In this example, seeing or not seeing the prejudice-reduction movies is manipulated in a random fashion. But the experimenters did not randomly cause some subjects to be male or female! The subjects came as they were. Therefore, to examine the effects of gender on response to the movies is nonexperimental analysis. In fact, it is a form of survey analysis—one characteristic of individuals (their gender) is examined to see if it is correlated with another characteristic (change in prejudice score). However, the subjects in this, and in most, experiments are not a sample of anything and do not constitute a population of any descriptive interest.

It might be interesting to know that American girls, or California girls, or female members of some other significant population are more open to persuasion on matters of prejudice. But we can't learn this from examination of kids at a few arbitrarily selected schools. So, when you read that men performed differently than did women in an experiment, know immediately that this is not an experimental result at all and probably is nothing more than a chance result based on a very small number of cases.

Experiments usually have only small numbers of subjects. It is typical for there to be only about 40 subjects in any group. Experimentalists are able to rely on small numbers precisely because no analysis is needed to control potential sources of spuriousness as is necessary with survey data, for example. Since, in a true experiment, everything besides the independent variable is controlled, only a small number of cases is needed to establish the statistical significance of an observed difference on the dependent variable. But, when nonexperimental findings are reported from such a database, they not only typically rest on few cases, but they have not been subjected to statistical analysis—perhaps the boys differed from the girls in other ways that account for the difference. Thus, such findings not only are limited to an accidental population of no interest, but they have not been subjected to even the most obvious tests for spuriousness.

External Validity

Many criticisms of the experimental method claim that the results obtained in such studies cannot be generalized to the "real world," which is beyond experimental control and manipulation. At issue here is the extent to which experiments have *external validity*.

The **external validity** of an experiment is the extent to which its findings can be generalized.

There are two basic types of generalization: statistical and theoretical (Meeker and Leik, 1995).

Statistical generalization involves successfully inferring that observations based on a sample apply to the unobserved members of that same population.

Theoretical generalization involves increasing the scope (generality) of tests of a theory by applying the theory to a variety of settings.

Experiments differ considerably in their capacity to sustain each type of generalization, depending on their setting. Obviously, experiments conducted as part of a survey can be generalized statistically—as can any survey finding. But there is little reason to suppose that statistical generalizations can be drawn from most laboratory experiments. This also applies to many field experiments, although for different reasons.

External Validity of Laboratory Experiments

Few laboratory experiments are based on a sample of any known population of interest. Subjects are volunteers, and their performance in an experiment cannot be assumed to describe that of anyone else. Although it usually is assumed that the main experimental effect would be found using other populations, it is not assumed that actual measurements on the dependent variable will be similar from one set of subjects to the next. That is, a set of movies may cause prejudice scores to decline in a whole series of experiments, but the actual average prejudice scores of groups can be expected to differ greatly from one set of subjects to another. This is why the fallacy of unmanipulated causes is so serious. It falsely assumes that descriptive findings can be generalized statistically to significant groups—for example, to male, American college students on the basis of observations of 35 male students recruited from Introductory Sociology at Southern Methodist University.

However, laboratory experiments are well designed to support theoretical generalizations. Virtually all laboratory experiments are designed to test a theory or theories, and their results generalize to the theory, increasing the array of situations in which the theory has held (or failed).

As an example, recall Malinowski's theory of magic: *Within any human group, magical and superstitious practices will be focused on those important activities over which people have the least control.* In Chapter 2, this theory was applied to baseball as follows: Baseball players have a great deal of control over their fielding and throwing. Errors are quite rare and the average player will succeed in making fielding plays correctly at least 95 percent of the time. Hitting is something else. The best hitters fail twice out of every three times at bat. Thus, it can be deduced from Malinowski's theory of magic that *baseball players will engage in many more superstitious practices about hitting than about fielding.*

But Malinowski's theory also can yield predictions about behavior in a laboratory. Suppose we recruit a number of subjects and ask each to play a computer game. We make this an important activity by offering substantial cash payoffs for points scored. Then, to create our independent variable, we rig the game so that various recurrent problems differ greatly in terms of how well players can perform them. That is, some required maneuvers are easy and players always succeed, while some maneuvers are very tricky and the rate of success is very low. Malinowski's theory tells us that, if we observe each subject carefully, *the extent to which subjects exhibit superstitious or magical practices will vary across maneuvers according to the rate of success.* Because such behavior may be rather subtle, it might be best to have each subject videotaped (perhaps through a one-way window) so that his or her behavior can be analyzed carefully and rated by independent judges. This strategy would link the experimental method to content analysis techniques, as will be discussed in detail in the next chapter.

Were such an experiment conducted and if the results supported the hypothesis, the scope of applications of Malinowski's theory would be increased, thereby adding to the generalization of the theory. One might claim that the lab situation is artificial in comparison with baseball, but the point seems irrelevant in that both hypotheses are valid applications of the theory and are of equal weight should they not be supported.

Another way to put this is that experiments do not generalize through the wall of the laboratory, but through its ceiling—to higher levels of abstraction. They do not necessarily generalize statistically in the sense of applying to other groups outside the lab, but they do generalize to the abstract theories from which their hypotheses are derived.

External Validity of Field Experiments

It might seem that these limits concerning statistical generalization would not apply to field experiments. But this is not necessarily so. Too often, the results of a field experiment are invalid because the experimental situation was not equivalent to the situation it was meant to assess.

Suppose the government is considering a program that would guarantee all adults a certain income. People without income or whose income fell below a certain amount would simply receive government checks without having to apply for welfare, thereby eliminating the huge administrative costs of the many welfare programs. Rather than seek aid for families with dependent children, food stamps, Medicaid, subsidized or public housing, and the like by showing that they met various qualifications, people would qualify simply by having insufficient income. Their checks would be for the difference between their current income (including no income at all) and some guaranteed minimum income. All people would need to do to qualify would be to report their current income.

Opposition to such a program probably would include the claim that it would greatly increase "welfare" payments because many people would stop working or would work less often and fewer hours, being content to live off the guaranteed income. Defenders of the proposal probably would argue that most people would prefer to be self-sufficient and to maximize their income; therefore, only the truly needy would rely on their guaranteed income.

To settle the dispute, a field experiment is designed. In three different cities, a census is conducted to identify employed persons earning less than the proposed guaranteed income. Of these, half are selected randomly to participate in a two-year trial of guaranteed income, and half are excluded from the program. After two years, comparisons between the two groups show that, during the first year, people included in the program did work fewer hours than did those excluded by the program, but, in the second year, the differences all but disappeared. Does this show that a guaranteed income will not cause people to work less and be more dependent on government subsidies? Absolutely not! Why? Because the experiment differs from the real program in critical ways.

Every person in the experimental group must know the program is only an experiment and that his or her income is going to be augmented only for two years. All of them probably remain concerned about how they will support themselves after the experiment ends. If they take two years off to enjoy their "win," they will have a harder time getting work when the experiment ends. That they did reduce their hours during the first year suggests a cautious effort to take as much advantage of the program as possible. In contrast, if the real program were put into effect, everyone would know they were entitled to a certain standard of living—not just for two years, but for life—whether they worked or not. For many people, working would have no significant economic benefits. It seems likely that the responses to the real program would be very different from responses to the experiment.

The disjuncture between the experimental situation and the "real world" has been demonstrated many times in field experiments devoted to new educational techniques. A careful survey of the studies published over the past 30 years reveals an amazing pattern—almost everything succeeds! Moreover, for many studies that found a specific technique improved student performance, there are other studies showing that a technique based on quite contrary assumptions worked, too. The reason is that the actual independent variables in these studies were not techniques, but teachers. Let's see how this happens.

Proponents of a new teaching technique often will arrange to demonstrate it by randomly selecting several schools or classes within a particular school system to constitute the experimental group, leaving the other schools to serve as a control group. Sometimes, the new technique in the experimental group classes is actually performed by those who developed the technique. When this is the case, the students in the experimental schools not only encounter a new educational technique but sometimes also far better-qualified teachers. As the developers of the technique, these teachers are extremely motivated—they really want the new method to work. Even when the innovators don't step in and take over the actual teaching, those doing the teaching have been given substantial recent training and usually are very enthusiastic about the new method. In contrast, the old method gets taught in the same old way by teachers who have had no special recent training in its use and who have not been fired up with new enthusiasm. It would be surprising if the students in the experimental group didn't do significantly better. But, once adopted, the new method soon is an old method and levels of teacher enthusiasm return to normal. And nothing was gained.

STUDYING CHANGE

Experiments are a very powerful method for studying the causes of changes in individuals and small groups, but experiments are limited as to time span. There is a huge literature consisting of laboratory experiments on attitude change (Cook, Fine, and House, 1995), for example, but most of the experiments are restricted to an elapsed time of no more than an hour or two. The six-month duration of the hypothetical study of prejudice reduction is very long as experiments go. Even major field experiments such as those designed to evaluate income-maintenance programs or new educational approaches seldom span more then two or three years. Moreover, because of the limited capacity of experiments to sustain empirical generalizations, it would be inappropriate to compare the descriptive results of similar experiments conducted over many years in order to estimate changes.

Social scientists who study social change seldom cite any experimental research.

ETHICAL CONCERNS

The huge field experiment to test the Salk polio vaccine injected half of the subjects with a placebo known to offer no protection against the disease. The experimenters did this in order to get an accurate baseline against which to compare the results found for the experimental group. Recall, however, that a lot more kids got polio in the control group than in the experimental groups and that four of them died. Most of these control group cases, and probably all of the deaths, could have been avoided had everyone received the vaccine—had there been no control group. But then, of course, it would not have been clear whether or not the drug worked. Some kids in the experimental group got polio, too. Without a control group, it would have been impossible to say whether the number of those children was lower than it would have been had they not been vaccinated.

Here is revealed, at its most agonizing level of severity, the fundamental ethical paradox raised by the experimental method: the potential benefits of very powerful evidence concerning causation versus the need to deceive some people even to the extent of refusing them a potentially life-saving treatment.

Most social scientific experiments do not risk lives. But they often do withhold what many might define as beneficial effects. For example, kids assigned to be educated by an old method will end up less educated than they would have been had the new method not been withheld, if the new method turns out to be effective. In similar fashion, had it turned out to be the case that a period of financial support following a person's release from prison substantially reduced the proportion who ended up back in prison, those in the control group might have felt they had been discriminated against.

There is no simple resolution to such concerns. But there is a general consensus that it is acceptable to withhold something from a control group as long as it leaves them no worse off than they would have been had they not taken part in the experiment. That condition is met in each of our examples. But it should be pointed out that, sometimes, people are worse off because they were in an experimental group. For example, what if the Salk vaccine had turned out to have deadly side effects? These are extremely complex matters and explain why drug testing is subject to strict government rules.

A far more common ethical issue concerning social scientific experiments has to do with deception. Many experiments are based on a great deal of deception. That is, what people are told about what is going on is a lie. Milgram's subjects were not administering shocks to people in order to speed their learning just as Darley's and Latanè's subjects were not taking part in a discussion group.

Obviously, these studies could not have been done unless the subjects had been deceived. The problem of demand characteristics requires that experimenters always withhold some information about what's going on. Here, too, it comes down to weighing the potential increase in knowledge against the potential harm done to subjects.

However, one rather firm rule has arisen among laboratory experimenters, which helps minimize harm: When subjects have been deceived, they must be fully *debriefed* at the end of the experiment. Thus, before sending the subjects on their way, Darley and Latanè explained to each one the aims of the study and why it had been necessary to mislead them. Afterward, Milgram introduced each "teacher" subject to his "victim" and explained that there had been no shocks and that no one was harmed by the experiment.

CONCLUSION

When it comes to demonstrating causation, no method comes close to the experiment—when its basic design is fulfilled. In this chapter, we have let you see why this is so and alerted you to the prevalence of quasi-experiments. If the independent variable was not manipulated and the manipulation was not random as to the treatment of individual subjects, it was not an experiment.

REVIEW GLOSSARY

- An **experimental group** consists of those persons randomly assigned to be exposed to the experimental stimulus—to the independent variable.

- A **control group** consists of those persons randomly assigned to not be exposed to the experimental stimulus. The purpose of the control group is to provide a baseline against which to assess the effect of the experimental stimulus.

- As used in experiments, a **placebo** is a substitute for the independent variable and is used to permit variation in the independent variable without permitting variation in other factors linked to the independent variable. For example, in an experiment to test a drug, one treatment group gets the drug while the other gets a harmless substitute, but *both* groups get injected.

- A **double-blind experiment** is one in which neither the subjects nor the experimenters know whether or when the independent variable varies.

- In experiments, a **demand characteristic** is an aspect of the design that tips off subjects as to what they are expected to do. Subjects tend to be cooperative and, therefore, to respond to demand characteristics.

- In the context of an experiment, **history** refers to any event occurring simultaneously with the independent variable that might account for the observed changes.

- In experiments, **maturation** refers to the effects of subjects getting older (maturing) as time passes.

- **Testing effects** occur when the tests or measurements are themselves causative agents—whether by changing views or by influencing performance, as in the case of subjects becoming testwise.

- **Instrument effects** refer to differences produced by variations in the accuracy of multiple measurements.

- **Regression to the mean** refers to the tendency of extreme scores to move (or regress) over time toward the group average (or mean).

- **Selection biases** occur whenever people are assigned to the experimental or control groups (or to groups based on levels of the independent variable) in non-random ways.

- **Quasi-experiments** are studies posing as experiments, but which fail to meet the essential experimental standards of internal validity. (*Quasi* means "having superficial resemblance to.")

- **Internal validity** consists of eliminating all of the sources of bias. That means there must be at least two groups to compare (two levels of the independent variable including present and absent) and that assignment to groups must be random.

- **Subject "mortality"** refers to the loss of subjects over the duration of an experiment.

- The **fallacy of unmanipulated causes** involves mistaking the effects of individual *characteristics of subjects* on the dependent variable for experimental effects.

- The **external validity** of an experiment is the extent to which its findings can be generalized.

- **Statistical generalization** involves successfully inferring that observations based on a sample apply to the unobserved members of that same population.

- **Theoretical generalization** involves increasing the scope (generality) of tests of a theory by applying the theory to a variety of settings.

CHAPTER 11

Content Analysis and Other Unobtrusive Techniques

Chapter 1 noted that the natural sciences have an advantage over the social sciences because bacteria and molecules don't blush. Since human beings often do blush as a result of being observed, those conducting surveys, field studies, and experiments must take care to ensure that the act of observation does not alter the behavior of those being studied. But there are many opportunities to measure social scientific concepts that run no risk of influencing behavior because they do not involve direct observation. In fact, such measurements typically take place after the behavior has occurred (sometimes centuries later), and those whose behavior was measured didn't know they would be studied and perhaps never did find out. Such measurement is referred to as *unobtrusive* (Webb, Campbell, Schwartz, and Sechrest, 1981).

UNOBTRUSIVE MEASUREMENT

An **unobtrusive measure** is one that has no effect on the objects being studied—it is a measurement that does not intrude. Such measures sometimes are referred to as *nonreactive measures*—ones that produce no reaction.

Unobtrusive measurements are possible because humans typically leave traces of their behavior. During the past several decades, archaeologists have reconstructed an exquisitely detailed portrait of the Maya civilization that flourished from 200 B.C. through A.D. 900 in Central America (Schele and Freidel, 1990). But these scholars have never met an ancient Mayan. What they have observed are the Mayans' buildings (including their magnificent temples), their sculpture, and their graves. The final breakthrough came when an international group of scholars succeeded in deciphering Mayan writing (Coe, 1992). This immense body of scholarship is based entirely on these cultural artifacts.

CULTURAL ARTIFACTS

An artifact is any object made by human work, and a **cultural artifact** is any such object that informs us about the physical and/or mental life of some set of human beings. The word *object* is interpreted very broadly to include written, filmed, and recorded material.

Fortunately for social scientists, humans leave an immense wake of cultural artifacts. For example, the museums of Europe are crowded with suits of armor worn by medieval knights. An enterprising scholar was able to calculate that the average European knight was only about five feet tall by measuring several thousand of these suits of armor. Scholars have reconstructed General George Custer's "Last Stand" at Little Big Horn by using metal detectors to locate cartridge casings; these enabled them to chart each successive firing line as Custer's troops withdrew toward a hilltop.

It will be useful to consider several examples in greater detail.

Turning Wine into Communicants

For many years there has been a debate about how religious Europeans were a few centuries ago. Some argue that once-upon-a-time people in Europe were very active church members, far more active than today. Others disagree, arguing that Europeans never were very involved in church and that the so-called Age of Faith is a nostalgic myth. Then a French sociologist discovered very appropriate unobtrusive data on religious practice in France centuries ago.

For at least 50 years, Catholic sociologists in Europe, especially in France and Belgium, have devoted themselves to parish studies—to exploring the dynamics of religious life at the very local level. Eventually, some of them began to explore the history of various parishes, utilizing the elaborate records that have been kept by parish priests for many centuries. One day Jacques Toussaert (in Delumeau, 1977) discovered solid data on religious practice hidden in these documents: receipts for the purchase of communion wine. Until recently, in order to remain in good standing, all Roman Catholics were required to attend a communion service at least once a year, Easter being the preferred time to do so. A communion service briefly reenacts the "Last Supper" during which Jesus shared bread and wine with his disciples. Hence, groups of parishioners come to the front of the church where the priest gives each a sip of wine and a wafer of unleavened bread.

There are many grumbles in surviving letters, documents, and official reports from medieval and early modern times that only a small number of persons in any French community met this obligation, despite the fact that "everyone" was ostensibly a Catholic. But Toussaert now had a basis for estimating the actual attendance at communion services. That is, knowing the amount of wine purchased for the occasion, and having established amount per sip, he could simply divide to obtain the maximum number of people who attended. Dividing this number by the number of those eligible to attend, yields a rate of attendance. If we suppose that the priest drank some of the wine, or that some was left over, or that some people gulped rather than sipped, then attendance was lower than Toussaert's figures show. But, even assuming that every drop was sipped by communicants, Toussaert ended up finding low levels of attendance. At least in terms of taking communion, the Age of Faith did not exist in medieval France.

Exploring Occult America

Suppose you wished to study the popularity of occult (nonstandard) religious, mystical, and magical beliefs and practices in the United States and Canada. The occult would include such beliefs and practices as psychic readings, astrology, fortune-telling, meditation, and the like. You could do a survey, asking questions such as this one included in the 1994 General Social Survey of the United States:

Astrology—the study of star signs—has some scientific truth.

Definitely true	9%
Probably true	38%
Can't say	10%
Probably not true	24%
Definitely not true	19%

Another approach would be to examine cultural artifacts. The *Yellow Pages* of the telephone books are a very good source. The complete *Yellow Pages* for the United States and Canada now are available on CD-ROM from several different companies. The software included with these databases allows a user to search for listings under a variety of headings and to search for listings containing various key words. Using the 1994 edition, we searched for all listings under each of these headings: Aquarian, Aquarius, Astrology, Astrologer, Aura, Card Reading, Card Reader, Color Therapy, Crystals, ESP, Fortune, Fortune Teller, Horoscope, Madame, Meditate, Meditation, Meditator, Metaphysical, Mind Reader, Mystic, Mystical, New Age, Occult, Palmist, Palmistry, Palm Reading, Parapsychology, Parapsychologist, Psychic, Psychics, Reader, Readings, Seer, Tarot, Telepath, Telepathic, Yoga, Zen, and Zodiac.

The software creates complete lists for each subject heading, including the full name for the listing (such as Mildred's Astrology Center), the city and state in which the listing is located, and the phone number. Many listings came up under more than one heading. We eliminated all duplicates from the combined list. Some listings were irrelevant—an immense number of conventional businesses in Mystic, Connecticut, were included in the list for *mystic*, for example. These also were eliminated (when in doubt, we simply called the number).

At that point, we began counting the total number of listings for each city and each state. Then we divided by the total population to create a rate of occult listings per million population. Following are the total listings and the rates for two U.S. states and two Canadian provinces.

Washington

Business Name	City
Akasha Metaphysical Bookstore	Bellingham
Astrological Therapies	Mercer Island
Astrology Club	Malaga
Astrology Et Al Metaphysical	Seattle
Astrology Transformational	Seattle
Aum-Nee Crystals and Books	Tacoma
B J International Psychic	Clinton
Barbara, Palm Reader	Blaine
Betty Atwater Astrology	Gig Harbor
CDM Psychic Institute	Seattle
CDM Psychic Institute	Spokane
CDM Psychic Institute	Tacoma

CDM Psychic Institute and School	Everett
Crystal Wizard Metaphysical Bookstore	Lynnwood
Egyptian Mysteries Astrology	Tacoma
Inner Sound Metaphysical Bookstore	Goldendale
Joshua—Astrologer	Tacoma
Marie's New Age Books	Leavenworth
Meditation Station	Bellevue
Metaphysical Counseling Service	Seattle
Metaphysical Institute	Seattle
Mystic Way	Seattle
Mystical General Store	Vashon Island
Mystic Power Psychic Treatment	Yakima
Northwest Center—Tarot and Arts	Seattle
Open Door Metaphysical Bookstore	Spokane
Osho Meditation Retreat	Redmond
Psychic Reader	Bellevue
Ruth, Psychic Reader	Seattle
Siddha Yoga Meditation Center	Seattle
Transcendental Meditation	Bremerton
Transcendental Meditation	Silverdale
Transcendental Meditation	Spokane
Unity Metaphysical Bookstore	Port Angeles

Rate per million population = 6.8

Minnesota

Business Name	City
Astrology by Moonrabbit	Minneapolis
Circle of Light Psychic Reader	St. Louis Park
Dial-A-Meditation	Edina
Fatima Psychic	St. Paul
Meditation Center	Minneapolis
Mikal and Associates Psychic Readings	Minneapolis
Minnesota Zen Meditation Center	Minneapolis
New Age Bookstore	Winona
New Age Stones	Plymouth
Psychic Reading Shop	Forest Lake
Shannon's Psychic Shop and Tarot	Minneapolis
Shirley Strasburg—Astrologer	Minneapolis
Siddha Meditation Center	St. Paul
Tarot by Marlene DeLott	Minneapolis
Transcendental Meditation Program	St. Paul
Uptown Psychic Studio	Minneapolis
West Suburban Meditation	Hopkins
Yoga Meditation Center	Hopkins

Rate per million population = 4.1

British Columbia

Business Name	City
Aster Metaphysical Counselling	Chemainus
Avalon Metaphysical Center	Victoria
Dennis A Readings, Inc.	Nanaimo
Dial-A-Meditation	Victoria
Fortune Teller and Associate	Richmond
Hermaglo Psychic International	Vancouver
International Meditation Society	Victoria
Joseph Ip Astrology Consultant	Vancouver
New Age Holistic Health Center	Prince Rupert
Nirvana Modern Metaphysics	Terrace
Phoenix Metaphysical Books	Surrey
Sara, Psychic Reader	Vancouver
Transcendental Meditation Center	Vancouver
Transcendental Meditation Center	Victoria
Vancouver Psychic Society	Vancouver

Rate per million population = 4.6

Quebec

Business Name	City
Centre Astrology Du Quebec	Granby
Centre Astrology M Perras	St. Jean-Sur-Richel
Centre de Meditation Transcendantale	Montreal
Madame Simone	Montreal
Madame Jennie	St. Basile-le-Grand
Meditation Transcendantale	Hull
Meditation Transcendantale	Dollard-de-Ormeau
Meditation Transcendantale	Sherbrooke
Meditation Transcendantale	Trois-Rivieres
Mireille Tarot	Sherbrooke
Palmistry Centre	Westmount
Transcendental Meditation Centre	Dollard-de-Ormeau

Rate per million population = 1.8

These data from the *Yellow Pages* show that occult activities are more prevalent in Washington than in Minnesota and that British Columbia's rate is higher than Minnesota's, lower than nearby Washington's, but much higher than Quebec's. When occult practitioners obtained business telephones, they would not have been wondering if we were watching them. This measure was entirely unobtrusive.

Many studies based on cultural artifacts involve nothing more elaborate than counting and comparing—although, as in this example, comparisons often will require that the totals be converted into rates. However, such studies sometimes involve very detailed and elaborate coding schemes. This approach is known as *content analysis*.

CONTENT ANALYSIS

Over the past few years, there has been increasing concern about violence in the media—not only on television and in the movies, but in song lyrics, comic books, magazines, and even newspapers. This concern prompts many possible questions:

- Is this movie more violent than that one?
- Are the movies produced by one studio more violent than those from another?
- Has the amount of violence shown on TV been increasing?
- Is there more violence in the lyrics of rap music than in the lyrics of heavy metal?
- Have comics become more violent?

Many answers have been offered to questions such as these. Often, these answers are simply unsupported judgments, regarded as self-evident: "To watch network television these days is to be subjected to unending violence." Other answers recite many examples of the violence found in a particular set of communications in order to demonstrate its prevalence. But there is no way to tell whether these examples are typical or even frequent—perhaps they are the only examples. The questions listed above and others like them can be answered only by proportional facts, not subjective impressions, guesses, or examples. Thus, finding answers to such questions requires that we obtain accurate and systematic data. But, how can we obtain "data" by watching television, reading comic books, or listening to music?

Content analysis systematically transforms nonquantified verbal, visual, or textual material into quantitative data to which standard statistical analysis techniques may be applied.

Put another way, content analysis involves transforming qualitative material into data, in accord with Stephan Thernstrom's witticism that "facts are not born but made." However, while survey researchers, field researchers, and experimentalists take an active approach to eliciting the data they use, content analysts specialize in *coding* data they played no part in eliciting—data created by others for reasons having nothing to do with research. For example, content analysts don't write or produce movies. Nor do they ask movie directors or producers about their movies. Instead, they watch the movies and systematically note certain specific aspects. For example, they might record each instance of violence, thus making it possible to compare movies in terms of the number of such instances. They also might categorize each instance of violence by type: torture, maiming, death, etc. Or they might rank the instances on the basis of explicitness or of severity. The same thing could be done with TV shows, comics, novels, or song lyrics.

What content analysts create is a *set of codes* based on concepts and *rules for applying these codes*. Analysis consists of seeking correlations (or contrasts) among the measures produced by applying the codes: Is the size of the production budget correlated with the amount or severity of violent content—do movies tend to be more violent the more expensive they are? Does the sex ratio of the cast correlate with the amount of violence—the higher the percentage of male roles, the more violent the film?

RELIABILITY

Content analysis depends on materials that were not created for the purposes to which the social scientists wish to put them. Those who gave speeches, wrote letters, produced movies, composed lyrics, edited magazines, painted portraits, or otherwise created the materials used in a content analysis were entirely unaware of the categories and coding rules that subsequently were used to quantify their efforts. These categories and codes are *imposed* on the content of these materials, and it is always possible, therefore, that the categories distort the true meaning of the material.

While content analysis is an entirely *unobtrusive measure*, this does not mean that the material was created without self-consciousness on the part of its creators—most of this material was created for specific audiences and with specific intentions as to its effect on those audiences. To whom this material was addressed often is of primary interest to content analysts.

MANIFEST AND LATENT CONTENT

In addition to imposing categories on materials, content analysts often attempt to quantify subjective judgments. For example, was a particular remark meant to be taken as sarcasm? Such judgments often are rather subtle and depend on the *latent* rather than the *manifest* content of the material.

Manifest content refers to the *explicit*, clear, and perhaps superficial meaning of verbal, visual, or textual materials.

Latent content refers to "deeper" or *implicit* meanings.

When someone writes or says, "I'd love that," the manifest content is that the person welcomes whatever "that" refers to. But facial expression and/or tone of voice may suggest that the person really means, "I'd dislike that a lot"—the latent content of the material. Coders often must make very subjective judgments to distinguish latent content from manifest content.

These aspects of content analysis make reliability a significant problem. There are two primary sources of unreliability in content analysis, unreliable codes and unreliable coders.

RELIABLE CODES

All of the principles applied to survey questions in Chapter 7 also apply to content analysis coding. If the codes are too subjective or too vague, if the categories overlap or are not exhaustive, if the categories do not provide sufficient scope for variation, or if the codes are biased, then the results may be unreliable because no one can apply them consistently or correctly.

RELIABLE CODING

Given a reliable coding scheme, a primary source of unreliability involves coder errors or inconsistencies. The coder misperceives or misunderstands the material to be coded and miscodes it. The coder mistakenly scores a case as 2 on a variable when he or she meant to score it as 1—perhaps the coder is careless or working too fast. The coder applies rules inconsistently so that an incident involving only verbal abuse is sometimes scored as violent and sometimes not. The problem may be an inattentive coder or even a biased coder. For example, a coder might code an incident differently depending on whether it involved a man shouting at a woman or a woman shouting at a man.

The solution to these problems is to use several coders, each of whom codes the *same* material. Then their coding is compared. Each disagreement is carefully checked and the appropriate code assigned, if possible. In all instances of which we are aware, comparisons across coders have revealed disagreements. Consequently, we suggest that content analysis should never be accepted unless multiple coders have been used and a complete report made on intercoder reliability.

VALIDITY

It is far easier to achieve reliable data from content analysis than it is to ensure that the data are valid. Here concern centers on biases in the material itself as well as on the extent to which we may draw inferences about more general matters from the materials. The following example will clarify these concerns.

Suppose we wish to explore the extent to which people in Colonial New England (prior to the American Revolution) did or did not openly express their sexual feelings and interests. We are in possession of a large number of love letters as well as many diaries written in this era. We also can examine novels written by New England authors at that time.

So we create a set of codes to classify these materials. Upon examination of the coding scheme, it is clear that the variables measuring the expression of sexual feelings and interests have compelling face validity. When a letter writer expresses admiration for his or her correspondent's sexual organs, for example, this is not a case of latent content. So, satisfied that our coding will be valid, we employ multiple coders so that it also will be reliable.

That takes care of validity, right? Hardly. Our data could be extremely biased and, therefore, could yield an entirely invalid portrait of sexual expression in New England.

DEPOSIT BIAS

The first major source of bias has to do with selection. Who wrote these letters, diaries, and novels? Whoever they were, they were not a random sample of the population.

Deposit bias refers to circumstances in which only some portion of the pertinent material was originally *included* in the set of materials to be coded.

Illiterates do not write letters, diaries, or novels. In fact, most people in most eras do not. Consequently, our set of materials is far more limited in scope than the question we wish to answer. As the social historian **Edward Shorter** (1975:9–10) noted,

> *What kind of people, after all, tended to burst forth in love letters, write novels, or compose memoirs other than a tiny elite at the pinnacle of the social order. [Such written materials] represent the experience of perhaps 5 percent of the population.*

> *The vast bulk of writing about intimate experience comes from people who had very little in common with the classes in which we are interested, the other 95 percent of the population. . . . The gulf between upper middle-class life and the experience of the lower orders was enormous in times past. And the accounts of the one are not acceptable substitutes for descriptions of the other.*

Deposit bias extends far beyond the realm of historical materials. Suppose you wished to compare the religiousness of communities based on the amount of church advertising in local newspapers. Other things being equal, we might conclude that communities having more column inches of church advertising (proportionate to their population) were more religious than communities with fewer inches. But other things probably are not equal, and many factors could greatly bias how many church ads are purchased. The papers in some communities may give churches a substantial discount thus increasing both the number and the size of the ads. There may be marked denominational differences in attitudes toward church ads—some denominations may regard advertising as tacky. Since communities also differ greatly in their denominational profiles, this might be the primary cause of observed differences. Or it may be that churches tend to advertise only when they aren't attracting enough people, and hence a lot of church advertising reflects a lower level of community religiousness. These are biases that would occur in the deposit of the materials.

SURVIVAL BIAS

Survival bias refers to circumstances in which only some portion of the pertinent phenomena was *retained* in the set of materials to be coded.

During the 1960s, an incredible number of small newspapers sprang up to report on the local counterculture. They differed from the "straight" press in their advocacy of drug use, opposition to the draft, and interest in new rock bands. But, like the hippie counterculture they represented, their days were numbered. By the mid-1970s, most of these papers were long gone and generally forgotten. Suppose someone today wanted to base a study of this period on the contents of the counterculture press. It would be hard to find copies of many of these papers. And it would be very hard to know whether the copies that still survive are representative of the whole. In fact, they probably are not since it seems reasonable to suppose that the more successful papers—the ones that had larger circulations and survived for a longer period—would differ significantly from the less successful ones and would be far more likely to be available still.

In assessing the validity of content analysis data, it is vital to consider the impact of both deposit and survival biases. What's missing from the collection and why? However, the major source of invalidity in the use of content analysis has to do with attempting to generalize findings based on verbal, visual, or textual materials to human populations. To be more specific, a magazine article is text, not flesh and blood.

INVALID GENERALIZATIONS

The questions posed earlier about media violence are entirely appropriate for content analysis. Each is a question about verbal, visual, or textual materials—about movies, TV, comic books, and the like. In each instance, it is relatively clear what materials ought to be coded and how to draw valid conclusions from such data.

However, social scientists frequently must rely on content analysis when what they really want to know about is people. In the hypothetical example about sexual expression in New England, the real research question was about how Puritans behaved, not about what letters or diaries contained. The hope was that these textual materials would offer a valid portrait of real life. Sometimes, such generalizations are valid. Sometimes, they aren't. As an instructive example, let's pause and examine a classic study, long offered as a model of well-done content analysis.

Leo Lowenthal believed there had been a major shift in cultural values in the United States during the first half of the twentieth century. He thought that, at the beginning of the century, people greatly admired what he called "idols of production"—founders of large corporations (such as Henry Ford) or people (such as Thomas Edison) who invented the items such corporations produced. But then, Lowenthal hypothesized, public preferences began to shift so that the most admired people were "idols of consumption"—movie and sports stars, for example.

To test this hypothesis, Lowenthal coded biographies that appeared in popular magazines between 1901 and 1941. The results gave very strong support to his hypothesis (Lowenthal, 1944). Idols of production dominated magazine biographies prior to World War I. But, by the late 1920s, the pattern had shifted and biographies of idols of consumption (such as Babe Ruth, Clark Gable, and Clara Bow) far outnumbered those of business leaders.

Lowenthal's findings were widely publicized and sustained an immense number of critiques of popular culture. In fact, his study had considerable impact on the most influential book of the time, *The Lonely Crowd: A Study of the Changing American Character* (Riesman, Glazer, and Denney, 1950). Although Lowenthal's data showed only what magazine editors chose to print, it seemed entirely valid to assume that their choices reflected popular taste. But it wasn't so.

More than a decade later, **Fred I. Greenstein** (1968) discovered a series of neglected surveys of elementary and high school students, each of which asked students whom they would "most like to resemble." The studies were done as dissertation projects by people doing graduate work in education departments. The oldest dated from 1902 and included 2,333 students. Others were conducted in 1910, 1928, 1944, and 1958.

What these data showed is that few kids *ever* wanted to be like idols of production, but they overwhelmingly named idols of consumption—Babe Ruth beat out Henry Ford many times over. Magazines had changed. Kids hadn't. Those who wished to claim a shift in the American character "are able to do so only by idealizing the American past" (Greenstein, 1968:305).

SELECTING UNITS AND CASES

The conflicting results of the studies were due to the use of different units of analysis. Lowenthal's units were biographical stories in popular magazines. Greenstein's units were students. It is always problematic to generalize results based on one kind of unit of analysis to another variety—from aggregates to individuals, for example. This problem is especially acute in content analysis because so often it is used as a substitute for data on individuals. Lowenthal didn't really care about shifts in popular biographies—he wanted to demonstrate changes in the preferences of people. When the real focus is on people, content analysts proceed at their own peril and should always be very aware of the risk that their results will not generalize to people.

APPROPRIATE UNITS

It is far better when content analysis is used to test hypotheses concerning the units of analysis used in the research. As noted, content analysis of movies can answer questions such as "Are movies more violent today?" Thus, the selection of units of analysis for content analysis generally should be determined by the research question. Suppose you had the hypothesis that American business communication has become less formal. Business letters would seem to be an appropriate set of units, and each letter would be a case. The hypothesis that retailers have reduced their emphasis on price and now stress quality could be tested with ads in catalogues, in newspapers, or on television—each ad would be a case.

Sample Versus Census

The decision to sample or to examine all available cases depends upon how many cases there are. If you have decided to limit the cases to one newspaper, for example, it might be possible to code all ads of a particular kind for a period of years. However, no one could code all such ads for all American papers even for the past year, let alone all ads for many years. Thus, it usually is necessary to select a sample of cases. The principles of sampling presented in Chapter 4 apply to *any* units of analysis, not just to people. First, it is necessary to define the population or universe of cases to be sampled. Then, random procedures must be followed in selecting cases.

Suppose you wished to analyze data based on personal ads in newspapers in which people solicited introductions to members of the opposite sex. First, you would need to identify the universe—perhaps all American daily newspapers that have a special section devoted to such ads. Next, you would need to draw a sample of papers (perhaps weighting them according to their circulation). Then, you would need to draw a sample of issues. At that point, you could either include all ads published on those dates by each paper in the sample or select a sample from each paper. You would then have a stratified random sample of personal ads. The next step would be to begin coding.

Constructing Coding Schemes

A content analysis coding scheme is, in an important sense, a questionnaire that coders fill out on behalf of each case. As such, it must adhere to the fundamental principles of question construction presented in Chapter 7. As with survey interview schedules or questionnaires, pretesting is vital. So, after you have a draft of your coding scheme, try it out on a few cases. You undoubtedly will discover many things that need to be changed. Fortunately, coding and recoding a case does not make it unfit for inclusion in the final study. Pretest survey respondents may be sufficiently changed by having been interviewed that it is unwise to include them in the final sample, but a letter, ad, or movie is a fixed communication.

In creating and refining a coding scheme, remember that the important issue is what you are going to do with the data after they are coded. What are your hypotheses? What data will you need to test them—are you leaving out something important? If you hypothesize that men who run ads seeking to meet women will be more apt to mention appearance when describing their preferences in women, while women will be more apt to stress long-term relationships in their ads, you will need to remember to code gender. While that may seem like absurdly obvious advice, you would be shocked at how often even very experienced researchers omit something really vital when collecting data.

Asking what you will do with the data also will help you recognize the inclusion of superfluous material. It is typical for preliminary coding schemes to include too many variables and for many to be dropped during pretesting.

The need to use multiple coders has been discussed. It also is best if the coders do not know the hypotheses to be tested. For example, if they know that the investigator thinks that, compared with women, men will emphasize looks in their personal ads, many coders will tend to shade their judgments in that direction. Therefore, although coders will need training in order to apply the codes accurately, this training must not tip off the researcher's expectations about findings.

STUDYING CHANGE

Content analysis often examines social change because cultural artifacts tend to be deposited over time. Magazines and newspapers, for example, invite the search for trends—thus, Lowenthal measured changes in the biographies published in popular magazines. The study of American occultism could be transformed into a study of change merely by coding *Yellow Page* listings from past years. Many studies have examined shifts in the proportion of nonwhites in ads and in television shows. Others have charted the rapid increase in the percentage of book sales made up by romance novels.

Studies of change based on verbal, visual, or textual materials must, of course, ensure that biases are constant over the period being studied. If materials were more or less likely to be deposited at various periods, the observed changes might not have taken place—the same applies to differential probabilities of survival. In general, however, no special methods beyond those used to analyze data from one period in time are required to study change.

A series of studies based on content analyses have assessed the hypothesis that the American news media are becoming increasingly judgmental and biased in their coverage of presidential election campaigns. **Kiku Adatto** (1990) coded TV network news coverage of the 1968 and 1988 campaigns. In 1968, when presidential candidates were shown on the screen, the spoken words accompanying the image were the candidates' own words 84 percent of the time. Moreover, the average "sound bite" of uninterrupted speech by a candidate on television news lasted 42 seconds. But, during the 1988 election, the average uninterrupted sound bite was only 10 seconds long, and the candidate's actual words were seldom heard. For every minute a presidential candidate was allowed to speak on the news, the reporter covering him was broadcast for 6 minutes. Replication research found the 1988 pattern held in 1992 (Patterson, 1994).

The hypothesis was about what was shown on television, not about how people may have responded to it—hence, TV news tapes were the appropriate units of analysis. Timing sound bites and reporters' comments does not present problems of validity. The generalization from these data seems clear: TV no longer *covers* the campaign as candidates actually conduct it but, instead, *interprets* the campaign.

These results were confirmed by **Thomas E. Patterson** (1994), who found that presidential candidates' words are also being squeezed out of the print media. Patterson coded all front-page presidential election campaign stories that appeared in the *New York Times* from the 1960 election (Kennedy/Nixon) through 1992 (Clinton/Bush/Perot). In 1960, the average quote of a candidate's words was 14 lines. In 1992, it was 6 lines. In 1960, most campaign statements by candidates were interpreted as representing their policies. In 1992, more than 80 percent of the time campaign statements were interpreted on the basis of strategy—the *Times* writers interpreted campaigns as games to be won or lost, not as reflections of political philosophies.

In addition, Patterson coded 4,264 paragraphs randomly selected from stories about major party presidential nominees, published in *Time* and *Newsweek* magazines during the 1960–1992 period. Among his coding categories were whether the paragraph contained evaluative references and, if so, whether these were favorable or unfavorable to the candidate. In 1960, of all evaluative references to John F. Kennedy and Richard M. Nixon, 75 percent were positive. In 1992, only 40 percent of reporters' evaluative references to Bill Clinton and to George Bush were favorable. Patterson also coded the covers of both magazines. In 1960, neither magazine printed a negative campaign cover. For example, when *Time* ran a cover story on Kennedy, the cover merely said, "Candidate Kennedy," and the Nixon cover said, "Candidate Nixon." In 1992, *Time* covers included "Why Voters Don't Trust Clinton" and "Is Bill Clinton for Real?"

Here, too, the selection of units of analysis was appropriate and the coding categories clear. Consequently, Patterson is justified in concluding that leading print publications have shifted from reporting what the candidates say to featuring partisan interpretations of what's "really" going on.

ETHICAL CONCERNS

Typically, unobtrusive methods have no capacity to harm. The material reported is public and does not relate to specific individuals. Of course, if you steal private letters or company documents to code, you are committing ethical as well as legal violations. However, there exist borderline cases in which care must be taken to protect the confidentiality of transactions and the privacy of individuals. For example, studies often are based on coding medical records. The confidentiality of doctor-patient relationships must be preserved, and it must never be possible for any individual case to be identified. The same principles apply to research based on other sorts of individual records: arrest records, credit histories, academic transcripts, or personnel files.

Conclusion

Of all the research methods examined in this book, content analysis is the cheapest. Most studies are conducted in the library, using published documents. All that is required is a worthwhile research question and some effort. For example, the study of occult America discussed early in the chapter required only the purchase of an inexpensive CD-ROM (about $150). The rest was nothing more than clerical work. It seems surprising that social science journals aren't bursting with content analysis, but it remains the least utilized method of research. Perhaps that is because social scientists tend to focus on individuals, while content analysis is better suited to studies of aggregate units or of cultural artifacts.

At the beginning of the book, we promised to let you see how social science is done—and especially how it is done right. For some of you, this will be the beginning of a social science career. But most of you will go on to other pursuits—we hope that this course will make you more skeptical and informed consumers of social scientific reports in the news media. However you apply what you have learned—as social scientists or informed consumers—we hope you enjoyed reading this book even half as much as we enjoyed writing it.

Review Glossary

- An **unobtrusive measure** is one that has no effect on the objects being studied—it is a measurement that does not intrude. Such measures sometimes are referred to as *nonreactive measures*—ones that produce no reaction.

- An artifact is any object made by human work, and a **cultural artifact** is any such object that informs us about the physical and/or mental life of some set of human beings. The word *object* is interpreted very broadly to include written, filmed, and recorded material.

- **Content analysis** systematically transforms nonquantified verbal, visual, or textual material into quantitative data to which standard statistical analysis techniques may be applied.

- **Manifest content** refers to the *explicit*, clear, and perhaps superficial meaning of verbal, visual, or textual materials.

- **Latent content** refers to "deeper" or *implicit* meanings.

- **Deposit bias** refers to circumstances in which only some portion of the pertinent material was originally *included* in the set of materials to be coded.

- **Survival bias** refers to circumstances in which only some portion of the pertinent phenomena was *retained* in the set of materials to be coded.

Appendix:
Writing the Research Report

By Michael Corbett

By this point in your academic progress, you have no doubt been given information several times on how to write research reports. The rationale for this appendix is that it applies more specifically to the social sciences, whereas much of the previous information to which you have been exposed might not have been so helpful in this particular area.

ORGANIZATION OF THE REPORT

How do you organize a report on social research? What goes into this report? To some extent, the answers to these questions will vary somewhat from one social science to another, but the variation is primarily a matter of style and labeling rather than substance. For example, the final major section of a report in which the implications of the results are discussed might be labeled Discussion or it might be labeled Conclusions. Thus, there is some variation. Overall, however, a social research report contains the following components:

- Title page
- Abstract
- Introduction
- Literature review
- Methods
- Results
- Discussion
- References

A report might also contain an appendix at the end if it is needed. In general, place material in an appendix if its inclusion in the main body of the paper would distract from the paper, if the material is not essential to understanding the rest of the paper, or if the material would be of interest to only some of the audience for the paper. For example, researchers sometimes use abbreviated survey questions in the discussion of results and then include the full wording of the questions in an appendix.

TITLE PAGE

A title page presents the title of the report and the name(s) of the author(s). Beyond this, the title page might present other types of information required by the particular circumstances of the report. For example, if you are writing a paper for a course requirement, then you will probably include the name of the course and perhaps the name of the professor. If the report is a paper to be submitted for possible publication, include your institutional affiliation. If the report is a paper to be presented at a conference, specify the name, date, and location of the conference.

ABSTRACT

The abstract provides a brief (generally 100 to 200 words) description of the report. The abstract should give enough information about the report that potential readers can make a decision about whether it is relevant to their interests. If so, then they can go on to read the rest of the report. If it is not relevant, then readers will not need to waste time reading the entire report. Figure A.1 presents an example of one of my own abstracts.

Figure A.1   Example of an Abstract

ABSTRACT

Trends Among the Young: Political Issues and Identifications, 1972–1991

Michael Corbett

This research uses NORC surveys from 1972 to 1991 to examine changes in political party identifications, liberalism-conservatism identifications, and issue stands among citizens in the 20–29 age category. During this time, young people became more likely to label themselves as Republicans and conservatives. Young people remained fairly stable in their views on certain issues (e.g., political tolerance, social welfare), but they became more conservative on some issues—especially crime issues. At the same time, they became more liberal on racial and sexual equality issues. Party identifications have not become more linked to issue stands, but the linkage between party identification and presidential voting became stronger from 1972 to 1991 among the young.

INTRODUCTION

The introduction sets the stage for the rest of the paper by presenting the research problem, indicating its background, and describing its importance. Some writers (e.g., Bem, 1981) have described the hourglass shape of a report. In this hourglass shape, the introduction to the paper begins with a very broad perspective. The introduction then proceeds to narrow this perspective as it reveals the specific focus of the research it concerns. This focus becomes even more defined as the report progresses through the literature review, the methods section, and the presentation of results. Then, in the discussion of the results, the focus broadens again to consider the broader implications of the study. Figure A.2 shows this flow.

While the nature of the introduction should depend partly on the intended audience for the report, the introduction should always place the reported research in context. If you are studying the major themes of movies during different time periods, what does this have to do with anything? What is the big picture here, and where does your research fit into that big picture? It might be, for example, that you are investigating one piece of the question about the influence of the mass media on social norms. In this case, you might start your introduction with a statement such as "Some writers have argued that the mass media have a tremendous influence in molding the public opinion." You would proceed from this very broad statement, to narrow down the focus to your present research—for example, examining the treatment of divorce in movies over time and linking this with actual divorce rates in the United States.

Figure A.2   The Hourglass Shape of a Research Report

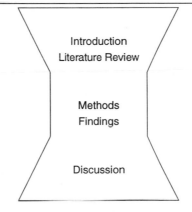

Introduction
Literature Review

Methods
Findings

Discussion

The introduction also needs to indicate the importance of the research topic in one way or another, depending upon the intended audience. The research might be important in solving a social problem, increasing knowledge in a particular area, helping to resolve contradictory findings in some area of research. Whatever the nature of the problem, the introduction should answer the question "Why is this research important?" In order to provide examples of opening statements that start out with the very broad part of the hourglass shape and also indicate the importance of the area of research, I just pulled from my bookshelf the March 1994 issue of the *Social Science Quarterly*. Let's look at some of the beginnings of several of these articles from the race and ethnicity section.

- "During the 1970s and 1980s the United States experienced substantial growth in undocumented immigration. Conservative estimates suggest that as many as three to four million undocumented immigrants were living in the United States in 1980" (Sorensen and Bean, 1994:1).

- "The Chicano population represents one of the fastest growing ethnic groups in the United States" (Saenz and Anderson, 1994:37).

- "If we wish to understand reasons for the severely disadvantaged status of some minority and majority population groups in America's cities, it is necessary to explain both who obtains jobs and who obtains good jobs" (Madamba and De Jong, 1994:53).

Each of these articles begins with a broad statement that indicates the importance of some area. Then the topic is narrowed down to the actual research undertaken. For example, Sorensen and Bean (1994) focus on the effects of the 1986 Immigration Reform and Control Act on the wages of recent Mexican immigrant groups, Mexican immigrants who arrived earlier, and Mexican Americans.

LITERATURE REVIEW

What do we already know about this topic of research? What research has already been done? What were the results? What are the gaps in the research? The literature review section puts the research into theoretical context so that we can see how it fits in with other studies—and why this study was needed.

The literature review summarizes the major points from the key studies relevant to the area under investigation. This review needs to be comprehensive in the sense that it covers all the major points that are relevant to your research. You need to find all the major studies that have significant findings for your specific area of research. Do not think in terms of finding a certain number of references—it is a big mistake to approach the literature review with the idea that you will get information on, say, 10 references and then stop. That approach could lead to serious gaps in your research. For example, if you reviewed the research on the effects of gender on social and political attitudes but missed an important study demonstrating that age is an important intervening variable, your report would have a problem.

While the literature review needs to be comprehensive, it also needs to be concise. Thus, you do not need to summarize every point from every study ever done in the area. Summarize the key points from the major studies that are relevant to what you are doing. Although there is variation from one professional journal to another, the literature review typically takes up about one to three printed pages in an article in a professional journal.

At the close of the literature review, you should have narrowed the focus down to your particular research. At this point, state (or restate) your specific research hypotheses or objectives. Further, it is a good idea to give a very general overview of your research.

Let's demonstrate the beginning and middle of the hourglass shape of the report—from introduction through the end of the literature review section—with the structure of an article titled "Reconsidering Race Differences in Abortion Attitudes" by John Lynxwiler and David Gay (1994). Lynxwiler and Gay began the article with a broad discussion of determinants of abortion attitudes and then narrowed the focus somewhat to differences between whites and blacks in attitudes toward abortion—blacks oppose abortion more than whites do. They then reviewed the literature, which demonstrated that blacks opposed abortion more than whites but that—depending upon which control variables were used—there might or might not be a relationship between race and abortion attitudes while controlling for certain attitudinal and socioeconomic background characteristics. Lynxwiler and Gay then briefly reviewed the literature, which supported the idea that women of childbearing years viewed abortion as an instrumental issue. Lynxwiler and Gay then proposed that, in comparing the abortion attitudes of white and black women, it is important to know whether those women are of childbearing age. Then the authors narrowed the focus (the middle of the hourglass shape of the report) to their particular research problem. They hypothesized that (controlling for the attitudinal and demographic characteristics used in past research) there would be no difference in abortion attitudes between white and black women of childbearing age—but there would be a difference in abortion attitudes between white and black women who were not of childbearing age.

METHODS

The methods section tells the reader how, when, where, by whom, and on whom or what the research was done. This section needs to be so complete that someone else could replicate your research—someone else could do the research the same way that you did it. How did we go about collecting the data? When were the data collected? Where were the data collected? How were the cases selected? What methods were used to analyze the data? The areas included in the methods section can vary a great deal depending upon the kind of research involved. For example, there would be substantial differences among the methods sections of the following reports:

- a study doing secondary analysis of a well-known national survey such as the NORC GSS survey
- a content analysis of newspaper editorials in major newspapers for different time periods

Despite the variation in what is needed in the methods section, let's look at some areas that usually need to be covered.

General Design of the Research.   You need to specify the general method of data collection (survey research, field research, etc.) and—depending upon the particular situation—the overall research design. In some cases, this can be handled very simply (e.g., secondary analysis of NORC GSS survey data), but other cases (e.g., a complicated experiment) might require greater explanation. In general, you need to present in this section any information that the reader might need in order to grasp the basic design of the research. In certain kinds of research (e.g., field research), the writer should describe the general setting and procedures at this point.

Selection of the Cases.   What cases were used in the research, and how were they selected? Often, the cases we use in social research are individual people, but this is not always the case. We might do research on cities, states, nations, groups, books, newspaper editorials, organizations, and so on. No matter what kind of cases we are using, we need to specify how these cases were selected. Again, the amount of detail that we need to include will vary depending upon the situation. If we are using the 1993 NORC GSS survey, we don't need to go into detail about sample selection because the general procedures of NORC are well known and fairly standardized. If you were doing your own survey, you would need to go into more detail about sample selection, response rates, and so on. If you were doing an experiment, you would need to specify how you selected the subjects.

Operationalization and Data Collection.   How did you operationally define your concepts, and how did you collect the data needed for this measurement process? For example, if you were investigating the relationship between personal trust and racism, how did you operationally define these concepts? If you used a questionnaire, you would need to specify the questions that you used—although the full wording of these questions might be placed in an appendix—and how, where, and when the survey was done. If you were doing an experiment, how did you operationalize the variables? In an experiment, for example, how would you operationalize the concept of personal trust? If you were doing field research, in what conditions was it done and how did you make systematic observations related to your concepts?

We often need to modify our data in various ways before doing the analysis, and the modifications that are relevant should be described in the methods section. For example, let's say that you compute a composite racism scale for each person in a survey based on answers to six racism questions. You need to specify the basic procedure by which this was done:

> For each of the six questions, we coded the racist response as a 1 and the nonracist response as a 0. Then we added the six scores together for each person to obtain a composite measure of racism with a possible range of 0 (lowest possible) to 6 (highest possible). We excluded respondents who did not answer all six questions.

Statistical Techniques.   If you are using conventional, straightforward methods of analysis (e.g., frequency distributions, contingency tables), there is no need to say anything here about such methods. However, if you are using some more complex method (e.g., factor analysis) or an unconventional method, then you should discuss it to some extent—depending upon how well known the statistical technique is—in the methods section.

RESULTS

The results section presents the findings of the research. In addition to including text, this section might include tables and graphic figures. Sometimes, a study produces a great deal more information than can be presented and discussed. In this situation, remember the main points of the research and the findings and focus the discussion on these. Do not ramble in the results section—keep the research problem in mind and focus the presentation of results so that they are directly relevant to the goals of the research. Some writers argue that you should present your most important findings first, but others take the opposite approach. Whichever approach you take, make sure to present the results that you needed for your research goal, and, in most situations, exclude results that are not relevant.

If you are doing quantitative research, you might have situations in which you have generated hundreds of tables or correlation coefficients during your statistical analysis. Let's say that you used contingency tables to examine relationships between 15 political attitudes and 8 religious attitudes while controlling for 4 socioeconomic background characteristics. This would produce 480 tables, and there's no practical way you can present all of these tables. Thus, you would need to focus on the most important results, present summary tables for somewhat less important results, and simply omit tables for those situations in which the results were unimportant.

In presenting quantitative results, you need to present any measures of association and significance test results that are appropriate. Sometimes, the measures of association and significance levels are the only kinds of statistical information that you need to present. In other situations, you might need to present other information such as descriptive statistics (e.g., averages, standard deviation), frequency distributions, or contingency tables.

If you are presenting qualitative data (e.g., informal observations or conversations), these data should also be presented in such a way that they can be used to reach conclusions about the research question under investigation. Everything presented in the results section should consist of findings that are relevant to the goal of the research—regardless of whether the findings are in accord with the expectations.

In any kind of research, we must let the chips fall where they may—we must present the results regardless of whether these were the results we had expected or hoped to find. One side effect of the scientific method is that we do sometimes find that our perceptions of reality are not correct. What do we do then? No one argues that we should try to fudge on the results, but there is some disagreement about the utility of revealing our lack of foresight. That is, some take the position that, if the results don't support the expectations, change the expectations before writing the report. This reminds me somewhat of the old joke about a man in a restaurant who apologetically tells the server that he has just enough money to pay the bill but no money left over for a tip. The server responds, "That's okay, sir. I'll just refigure the bill." While there's no doubt that some social researchers do refigure their expectations if the results don't turn out "right," there is no allowance for any refiguring on the results themselves.

DISCUSSION

The discussion section often begins—or sometimes ends—with a summary of the research problem and the results. Depending upon the complexity of the research, the length of this summary might range from a few sentences to one or two pages. By this point in your report, the reader may have lost track of the forest because of all the trees. Thus, the summary helps to put the overall research objectives and findings back into perspective.

We reach the bottom part of the hourglass shape of the report, and the focus expands again—the discussion turns to the broader implications of the results. What are the consequences of the research for the state of knowledge in this area? Has the research supported or refuted certain theories? If the results are not in accord with the specified hypotheses, what new hypotheses or theories might be pursued? What new research is needed? Do the research findings have implications for society?

The discussion section usually indicates both the successes and the failures of the research. What has the research accomplished and what didn't it accomplish? Were there unusual conditions that might have contributed to the nature of the results? Does the research need to be replicated under other conditions? Were there limitations on the research that need to be taken into account when interpreting the results? In short, if someone else were going to do research in this area, how should they do the research differently?

REFERENCES

The references section presents an alphabetic listing of the books, articles, or other materials used in writing the report. Ordinarily, you would list only those materials that you cited in the report somewhere. In some cases, however, you might present a bibliography instead of a list of references. The difference is that references include only those materials that you cite in the report, whereas bibliographies can include other materials as well. A bibliography, for example, might include materials that you read for background on the research topic but did not need to cite in the main body of the paper. Also, a bibliography might include additional materials that you think other people would be interested in reading—regardless of whether you had read them yourself.

There are many different styles for references, and new variations are appearing all the time. Thus, when you write a paper, you need to find out what style is required for that particular paper. In almost all cases, the same information is presented, but the way in which the information is presented can vary. A book reference contains the name(s) of the author(s), the title of the book, the year when the book was published, the city where the book was published, and the name of the publisher. Note that the two different styles that follow contain almost exactly the same information—the chief difference in information is the name difference— but the order of presentation differs.

Becker, Howard S. 1986. *Writing for Social Scientists: How to Start and Finish Your Thesis, Book, or Article.* Chicago: University of Chicago Press.

Becker, H. S. *Writing for social scientists: how to start and finish your thesis, book, or article.* Chicago: University of Chicago Press, 1986.

WRITING STYLE

In addition to thinking about what you present in a report, you need to pay attention to how you present it. Let's look at several ideas about writing style.

CLARITY AND PRECISION

The major goal in writing a research report is to present your ideas clearly and precisely. Most of the comments here apply to this goal in one way or another, but you should keep this in mind as the major goal throughout the process of writing the paper. Ask yourself this question: Will this paper give the reader a clear and accurate picture of my study? In doing this, of course, you need to consider who the reader will be. That is, what is the audience for which you are writing? Writing that is clear and precise for one audience (e.g., other social science students in your area) might seem like Greek to a different type of audience.

If you can write in an interesting and creative style, that is a bonus. However, do not sacrifice clarity and precision in order to make the report interesting or creative. Do not add creative touches that will distract from the main points of the report. Also, consider the likely reaction of your audience to anything that you have added in order to make the report more interesting or creative. One of the best ways to make a report both clear and interesting is to provide interesting examples that clarify the points of your research.

SENTENCE STRUCTURE

Keep two ideas in mind concerning the length of sentences:

- Do not use extremely long sentences.

- Use a mixture of short, medium, and long sentences.

Extremely long sentences are likely to cause confusion, and a long series of very short sentences might seem choppy. By using a mixture of different sentence lengths, you add some variety to the writing style and keep it from becoming monotonous.

If a sentence is excessively long, it might need to be broken into two or more sentences or it might be shortened by eliminating unnecessary words or phrases. Consider the following sentence: Many times when you are writing and you have excessively long sentences, you should give careful consideration to the question of whether there are some unnecessary or superfluous words or phrases in particular sentences that you might very well eliminate without sacrificing any of the meaning contained in the sentence. The preceding sentence is a deliberate example of this problem. Let's rephrase it: Shorten long sentences by eliminating unnecessary words or phrases.

Another matter concerning sentence structure is the issue of using the active rather than the passive voice. In general, you should use the active voice. Consider the following two sentences:

Passive: The issue of neighborhood crime was raised by Maria Gonzalez during the city council meeting.

Active: Maria Gonzalez raised the issue of neighborhood crime during the city council meeting.

Although this is a matter of opinion, most social scientists seem to prefer the second sentence—the active voice version. Many argue that the passive voice is more likely to cause confusion and that sentences using the passive voice are less interesting.

TERMINOLOGY

There are several important cautions concerning the selection of words to use in reports.

First, avoid the use of slang terms. Slang terms come and go, and sometimes the meaning of a particular term changes over time. Also, slang terms often have an imprecise meaning. Further, a particular generation often has its own slang terms that other generations might not understand. Thus, the use of slang terms makes it more difficult to achieve cross-generational understanding.

Second, avoid inappropriate jargon. In many situations, we can communicate our ideas very clearly, precisely, and efficiently by using specialized jargon. However, when you use jargon (e.g., social distance, anomie, privatism, regression toward the mean, recidivism), make sure that it is appropriate and that your audience will understand what you mean. Also, make sure that you yourself understand the jargon that you are using. I recently read a letter by a student who, in the same sentence, accused his professor of being elitist and egalitarian!

Third, avoid abbreviations that are not generally understood. In general, you should spell out the whole names of things unless you're referring to something that is very well known by its initials or an abbreviation. There would be no problem with the following sentence: CBS, ABC, and NBC all covered the presidential press conference. It would not be necessary, for example, to spell out Columbia Broadcasting System. On the other hand, what comes to your mind when you hear the initials ADA? You might think of Americans with Disabilities Act, Americans for Democratic Action, American Dental Association, or American Dairy Association.

Fourth, many social scientists still consider it inappropriate to use first-person pronouns (I, my, me, mine, etc.). The underlying idea here is that the use of first person implies subjectivity, and objectivity and neutrality are desirable in doing social research. These days, however, there are many social scientists who not only accept but encourage some usage of personal pronouns.

USING THE LIBRARY

You need to use the library in order to do a literature review for your research topic. In the social sciences, we rely primarily on books and articles in professional journals (e.g., *American Political Science Review, American Journal of Sociology, Journal of Social Welfare, Social Problems, Criminology*). Additionally, we need to consider magazine articles, newspaper articles, theses, conference proceedings, and perhaps films or other visual materials. As time goes by, we might also need to consider sources that are exclusively computer based—such as information, viewpoints, and debates that travel over the "information superhighway" via Internet, CompuServe, America Online, Prodigy, or some other electronic network.

Let's say that you are doing a research project on racism. How would you go about finding information on racism in the library? You could start by asking a reference librarian. However, it is likely that the reference librarian would get you started on this search by showing you one or more of the types of methods discussed here. Instead of using the reference librarian as a starting point, use the reference librarian when you are stuck and really need help.

You could, of course, start your literature search in the card catalog. We won't go into detail here about the card catalog because it is not likely that you will be using one and—if you do need to use one—you probably already know how to use it. Typically, card catalogs are not used much in most universities and colleges these days.

When you begin a search, you might be overwhelmed by the amount of information available. For this reason, it is a good idea to start with the most recent literature and work backward. In this way, you are in a better position to discover which research is still relevant, which research has been further supported through replication, which research findings were modified as new research was done, and which research ultimately led to a dead end.

INDEXES, ABSTRACTS, AND BIBLIOGRAPHIES

In order to do a search for a particular topic in the library, you might use one of the indexes, abstracts, or bibliographies. Some of these are general (e.g., the *Reader's Guide to Periodical Literature*), but some are more oriented toward the social sciences. Here is a sample of some of these guides:

- *Social Science Index*
- *Social Science Citation Index*
- *Index to Social Sciences & Humanities Proceedings*
- *Sociological Abstracts*
- *Criminal Justice Abstracts*
- *ABC Pol Sci: A Bibliography of Contents*
- *Current Contents/Social and Behavioral Science*
- *Sage Public Administration Abstracts*
- *Psychological Abstracts*
- *Women's Studies Abstracts*

Let's take a closer look at two of these guides.

Social Science Index.   The *Social Science Index* is often a good place to start a search. It is a quarterly publication (also available in some libraries on CD-ROM) that contains information about articles published in about 300 journals related to the social sciences. This index is set up so that you can search it by either subject or author. The electronic format allows even greater flexibility because you can search by author, subject, and key words.

Social Science Citation Index.   The *Social Science Citation Index* is published three times a year (also available on CD-ROM) and includes information from about 2,000 journals related to the social sciences in one way or another. In addition to serving as an index, the *Social Science Citation Index* lists the places where a particular work has been cited by others.

For example, let's say that, in searching for articles on racism, you have found an article titled "A Comparison of Symbolic Racism Theory and Social Dominance Theory as Explanations for Racial Policy Attitudes" by Jim Sidanius, Erik Devereux, and Felicia Pratto. This article was published in the June 1992 edition of the *Journal of Social Psychology*. Along with this listing, the index will include citations of this work by other articles in the index.

The listing of citations of a particular article is helpful in two ways. First, it helps you to locate other articles that deal with a particular topic. If the racism article mentioned is cited in four other articles, it is likely that these four other articles will be helpful to you in doing research on racism. Second, the sheer number of citations for an article might be a rough indicator of the importance of the article. If an article is frequently cited, this might mean that it has had a great impact or it might mean something else—e.g., that many other researchers cite it as an example of flawed research or that the article is simply controversial in some way. However, if an article is not cited by any other researchers, that is at least a rough indication that it has not had much impact in the area of research it concerns.

COMPUTER SEARCHES

These days instead of dealing with printed indexes, you would probably do a computer search in order to find sources for your research topic. In general, computer searches could include any of the following types:

- a search for books and perhaps electronic sources within a particular library system—e.g., OPAC (Online Public Access Catalog)

- a search for articles—e.g., ERIC (Educational Resources Information Center), EXAC (Expanded Academic Index), or NABS (Newspaper Abstracts)

- a search for materials in a library network that, for example, ties together the libraries of several different colleges or universities in a city, state, or other area

- a search for materials over a much broader network such as the LIBS system, which can search many, many libraries over the Internet

- a search of materials on a self-contained database such as the *Social Science Index* on CD-ROM

Continuing with the example of racism as the research topic, I used my university's OPAC system to search for books related to racism. The results of this search varied depending upon whether I searched on the basis of subject or key words in the title. The search specifying "racism" as the subject resulted in a list of 346 books. Next, I used EXAC to search for articles (since 1988) on racism, and this produced a list of 957 articles. In order to demonstrate the type of information you can get from this index, I selected one article more or less randomly and obtained the complete information given by OPAC about this article. Figure A.3 presents this information in slightly modified form for presentation purposes. Note that it includes an abstract that can be extremely useful in deciding whether you need to obtain and read the entire article.

A further convenience of computer searches is that there is often a means of printing the results without having to take the time to write down the information. You might also be able to save the results into a computer file that you can search further and edit before printing. For example, if you are searching Harvard's library from California through the LIBS system, you might have a computer logging program that simply records everything on the screen. Later, you can edit this information before printing just the information you need.

Figure A.3   Example of Results (Modified for Presentation Purposes) from a Computerized Search of
Library Journal Articles Using EXAC (Expanded Academic Index)

Search Request: S = RACISM

BIBLIOGRAPHIC RECORD—565 of 957 Entries Found

Author:	McMahon, Anthony
Other Authors:	Allen-Meares, Paula
Title:	Is social work racist? A content analysis of recent literature.
Journal:	Social Work Nov 1992, v37, n6, p533(7)
ISSN:	0037-8046
Abstract:	Social work with minorities seems to be a marginal interest for social work profes-sionals. What social work efforts are made with minorities appear to be naive and superficial and fail to address their social context. These are some of the findings highlighted in a content analysis of recent literature on the social work profession. The naivete and superficiality of social work efforts with minorities spring from social workers' failure to distinguish between individual and institutional racism. A more advocative, organized and antiracist stance is thus sought from the profession to bridge the gap between reality and their ideals.

The resources of libraries today make it easier for you to obtain information about research topics. On the other hand, easy access to a greater amount of information doesn't necessarily make your research better. This access does provide the potential for better research, but it must be used in conjunction with thinking and planning.

FOR FURTHER READING

American Psychological Association. 1994. *Publication Manual of the American Psychological Association.* 4th ed. Washington, D.C.

Becker, Howard S. 1986. *Writing for Social Scientists: How to Start and Finish Your Thesis, Book, or Article.* Chicago: University of Chicago Press.

Leedy, Paul D. 1993. *Practical Research: Planning and Design.* 5th ed. New York: Macmillan.

Strunk, William, Jr., and E. B. White. 2000. *The Elements of Style.* 4th ed. New York: Allyn and Bacon.

References

Adatto, Kiku. 1990. *Sound Byte Democracy: Network Evening News Presidential Campaign Coverage, 1968 and 1988.* Cambridge, MA: John F. Kennedy School of Government, Harvard University.

Albrecht, Stan L., Bruce A. Chadwick, and David Alcorn. 1977. "Religiosity and Deviance: Application of an Attitude-Behavior Contingent Consistency Model." *Journal for the Scientific Study of Religion* 16:263–274.

Allport, Gordon. 1958. *The Nature of Prejudice.* New York: Doubleday.

American Psychological Association. 1994. *Publication Manual of the American Psychological Association.* 4th ed. Washington, DC.

Anderson, Margo J. 1988. *The American Census: A Social History.* New Haven: Yale University Press.

Bainbridge, William Sims. 1996. *The Sociology of Religious Movements.* New York: Routledge.

———. 1982. "Shaker Demographics 1840–1900: An Example of the Use of U.S. Census Enumeration Schedules." *Journal for the Scientific Study of Religion* 21:352–365.

Barker, Eileen. 1984. *The Making of a Moonie.* Oxford: Basil Blackwell.

Becker, Howard S. 1986. *Writing for Social Scientists: How to Start and Finish Your Thesis, Book, or Article.* Chicago: University of Chicago Press.

———. 1953. "On Becoming a Marijuana User." *American Journal of Sociology* 59:235–242.

Bem, Darly J. 1981. "Writing the Research Report." In Louise H. Kidder, Selltiz, Wrightsman, and Cook's *Research Methods in Social Relations, Fourth Edition.* New York: Holt, Rinehart and Winston. 340–364.

Berelson, Bernard. 1978. "Prospects and Programs for Fertility Reduction: What? Where?" *Population and Development Review* 4:579–616.

Berezin, Mabel. 1994. "Cultural Form and Political Meaning: State-Subsidized Theater, Ideology, and the Language of Style in Fascist Italy." *American Journal of Sociology* 99:1237–1286.

Binder, Amy. 1993. "Constructing Racial Rhetoric: Media Depictions of Harm in Heavy Metal and Rap Music." *American Sociological Review* 58:753–767.

Blalock, Hubert M., Jr. 1979. *Social Statistics: Revised Second Edition*. New York: McGraw-Hill.

———. 1966. "The Identification Problem and Theory Building: The Case of Status Inconsistency." *American Sociological Review* 31:52–61.

Blumstein, Alfred, Jacqueline Cohen, and Richard Rosenfeld. 1991. "Trend and Deviation in Crime Rates: A Comparison of UCR and NCS Data for Burglary and Robbery." *Criminology* 29:237–263.

Bogardus, Emory S. 1924. *Fundamentals of Social Psychology*. New York: Appleton-Century.

Boudon, Raymond. 1965. "A Method of Linear Causal Analysis: Dependence Analysis." *American Sociological Review* 30:365–374.

Bradburn, Norman, and Seymore Sudman. 1979. *Improving Interview Method and Questionnaire Design*. San Francisco: Jossey-Bass.

Bromley, David G., and Anson D. Shupe, Jr. 1979. *"Moonies" in America: Cult, Church, and Crusade*. Beverly Hills, CA: Sage.

Broude, Gwen J., and Sarah J. Greene. 1983. "Cross-Cultural Codes on Husband-Wife Relationships." *Ethnology* 22:263–280.

———. 1976. "Cross-Cultural Codes on Twenty Sexual Attitudes and Practices." *Ethnology* 4:409–429.

Brownlee, K. Alexander. 1955. "Statistics of the 1954 Polio Vaccine Trials." *Journal of the American Statistical Association* 50:1005–1013.

Brustein, William. 1991. "The 'Red Menace' and the Rise of Italian Fascism." *American Sociological Review* 56:652–664.

Burkett, Steven R., and Mervin White. 1974. "Hellfire and Delinquency: Another Look." *Journal for the Scientific Study of Religion* 13:455–462.

Burstein, Paul. 1991. "Legal Mobilization as a Social Movement Tactic: The Struggle for Equal Employment Opportunity." *American Journal of Sociology*. 96:1201–1225.

Campbell, Donald T., and Julian C. Stanley. 1963. *Experimental and Quasi-Experimental Designs for Research*. Chicago: Rand McNally.

Carmines, Edward G., and Richard A. Zeller. 1979. *Reliability and Validity Assessment*. Beverly Hills, CA: Sage.

Carnap, Rudolf. 1953. "Testability and Meaning." In *Readings in the Philosphy of Science,* edited by Herbert Feigl and May Brodbeck. New York: Appleton-Century-Crofts.

Cary, Mark S. 1978. "Does Civil Inattention Exist in Pedestrian Passing?" *Journal of Personality and Social Psychology* 36:1185–1193.

Chagnon, Napoleon A. 1988. "Life Histories, Blood Revenge, and Warfare in a Tribunal Population." *Science* 239:985–992.

Coe, Michael D. 1992. *Breaking the Maya Code*. New York: Thames and Hudson.

Converse, Jean M. 1987. *Survey Research in the United States: Roots and Emergence, 1890–1960*. Berkeley: University of California Press.

Converse, Jean M., and Stanley Presser. 1986. *Survey Questions: Handcrafting the Standardized Questionnaire*. Newbury Park, CA: Sage.

Converse, Philip E. 1964. "The Nature of Belief Systems in Mass Publics." In *Ideology and Discontent*, edited by David Apter. New York: Free Press.

Cook, J., and N. Schoeps. 1985. "Program Response to Mail Surveys as a Function of Monetary Incentives." *Psychological Reports* 57:366.

Cook, Karen S., Gary Alan Fine, and James S. House, eds. 1995. *Sociological Perspectives on Social Psychology*. Needham Heights, MA: Allyn and Bacon.

Cronbach, L. J. 1951. "Coefficient Alpha and the Internal Structure of Tests." *Psychometrika* 16:297–334.

Cuomo, Mario M. 1983. *1933–1983—Never Again*. A Report to the National Governors' Association Task Force on the Homeless.

Cutright, Phillips, and Lowell Hargens. 1984. "The Threshold Hypothesis: Latin America 1950–1980." *Demography* 21:435–458.

Darley, John M., and Bibb Latane. 1968. "Bystander Intervention in Emergencies: Diffusionof Responsibility." *Journal of Personality and Social Psychology* 8:377–383.

Davis, Kingsley. 1945. "The World Demographic Transition." *Annals of the American Academy of Political and Social Sciences* 271:1–11.

DeLamater, John D. 1992. "Attitudes." In *Encyclopedia of Sociology*, edited by Edgar F. and Marie L. Borgatta (pp. 117–124). New York: Macmillan.

Delumeau, Jean. 1977. *Catholicism Between Luther and Voltaire*. Philadelphia: Westminster Press.

Dion, Karen. 1972. "Physical Attractiveness and Evaluations of Children's Transgressions." *Journal of Personality and Social Psychology* 24:207–213.

Duncan, Otis Dudley, R. P. Cuzzort, and Beverly Duncan. 1961. *Statistical Geography: Problems in Analyzing Areal Data*. New York: The Free Press.

Durkheim, Emile. 1897 [1951]. *Suicide: A Study in Sociology*. New York: The Free Press.

Eberstadt, Nicholas. 1995. *The Tyranny of Numbers: Measurement and Misrule*. Washington, DC: The AEI Press.

Ekstrand, Laurie E., and William A. Eckert. 1981. "The Impact of Candidate's Sex on Voter Choice." *Western Political Quarterly* 34:78–87.

Erchak, Gerald M., and Richard Rosenfeld. 1994. "Societal Isolation, Violent Norms, and Gender Relations: A Reexamination and Extension of Levison's Model of Wife Beating." *Cross-Cultural Research* 28:111–133.

Evans-Pritchard, Sir Edward. 1981. *A History of Anthropological Thought.* New York: Basic Books.

Farley, Reynolds, and William H. Frey. 1994. "Changes in the Segregation of Whites from Blacks During the 1980s: Small Steps Toward a More Integrated Society." *American Sociological Review* 59:23–45.

Festinger, Leon, Henry W. Riecken, and Stanley Schachter. 1956. *When Prophesy Fails.* New York: Harper and Row.

Finkel, Steven, Thomas M. Gutterbock, and Marian J. Borg. 1991. "Race-of-Interviewer Effects in a Preelection Poll: Virginia 1989." *Public Opinion Quarterly* 55:313–330.

Firebaugh, Glenn, and Frank D. Beck. 1994. "Does Economic Growth Benefit the Masses? Growth, Dependence, and Welfare in the Third World." *American Sociological Review* 59:631–635.

Fishbein, Martin, and Icek Ajzen. 1975. *Belief, Attitude, Intention, and Behavior.* Reading, MA: Addison-Wesley.

Freeman, Derek. 1983. *Margaret Mead and Samoa: The Making and Unmaking of an Anthropological Myth.* Cambridge, MA: Harvard University Press.

Frey, J. H. 1983. *Survey Research by Telephone.* Beverly Hills, CA: Sage.

Glenn, Norval D. 1977. *Cohort Analysis.* Beverly Hills, CA: Sage.

Goffman, Erving. 1963. *Behavior in Public Places.* New York: The Free Press.

Goldman, Marion S. 1981. *Gold Diggers and Silver Miners: Prostitution and Social Life on the Comstock Lode.* Ann Arbor: University of Michigan Press.

Goleman, Daniel. 1993. "Experts Advise a Healthy Skepticism About Polls." *New York Times,* Sept. 7, p. B5.

Gottfredson, Michael R., and Travis Hirschi. 1990. *A General Theory of Crime.* Stanford, CA: Stanford University Press.

Goyder, John. 1985. "Face-to-Face Interviews and Mailed Questionnaires: The Net Difference in Response Rate." *Public Opinion Quarterly* 49:234–252.

Greeley, Andrew M. 1989. *Religious Change in America.* Cambridge, MA: Harvard University Press.

Greenstein, Fred I. 1968. "New Light on Changing American Values: A Forgotten Body of Survey Data." In *Sociology and History: Methods,* edited by Seymour Martin Lipset and Richard Hofstadter (pp. 292–310). New York: Basic Books.

Hallam, Elizabeth M. 1986. *Domesday Book Through Nine Centuries.* London: Her Majesty's Stationery Office.

Hamberg, Eva M., and Thorleif Pettersson. 1994. "The Religious Market: Denominational Competition and Religious Participation in Contemporary Sweden." *Journal for the Scientific Study of Religion* 33:205–216.

Hanson, Robert C. 1958. "Evidence and Procedure Characteristics of 'Reliable' Propositions in Social Science." *American Journal of Sociology* 63:357–370.

Harris, Marvin. 1979. *Cultural Materialism: The Struggle for a Science of Culture*. New York: Random House.

Hatchett, Shirley, and Howard Schuman. 1975. "White Respondents and Race-of-Interviewer Effects." *Public Opinion Quarterly* 39:523–528.

Hempel, Carl G. 1952. *Fundamentals of Concept Formation in Empirical Science*. Chicago: University of Chicago Press.

Heyns, Barbara. 1978. *Summer Learning and the Effects of Schooling*. New York: Academic Press.

Higgins, Paul C., and Gary L. Albrecht. 1977. "Hellfire and Delinquency Revisited." *Social Forces* 55:952–958.

Hindelang, Michael, Travis Hirschi, and Joseph G. Weis. 1981. *Measuring Delinquency*. Beverly Hills, CA: Sage.

Hirschi, Travis. 1969. *Causes of Delinquency*. Berkeley: University of California Press.

Hirschi, Travis, and Hanan Selvin. 1967. *Delinquency Research*. New York: The Free Press.

Hirschi, Travis, and Rodney Stark. 1969. "Hellfire and Delinquency." *Social Problems* 17:202–213.

Hite, Shere. 1987. *Women and Love: A Cultural Revolution in Progress*. New York: Alfred A. Knopf.

Homans, George. 1967. *The Nature of Social Science*. New York: Harcourt, Brace & World.

———. 1950. *The Human Group*. New York: Harcourt, Brace & World.

Hovland, Carl, Arthur A. Lumsdaine, and Fred D. Sheffield. *Experiments in Mass Communications*. Princeton, NJ: Princeton University Press, 1949.

Hunter, Alfred A., and Margaret Denton. 1984. "Do Female Candidates 'Lose Votes'?: The Experience of Female Candidates in the 1979 and 1980 Canadian General Elections." *Canadian Review of Sociology and Anthropology* 21:395–406.

Iannaccone, Laurence R. 1994. "Why Strict Churches Are Strong." *American Journal of Sociology* 99:1180–1211.

Jacob, Herbert. 1985. *Using Published Data: Errors and Remedies*. Beverly Hills, CA: Sage.

Janus, Samuel S., and Cynthia L. Janus. 1993. *The Janus Report on Sexual Behavior*. John Wiley & Sons.

Jasso, Guillermina, and Peter H. Rossi. 1977. "Distributive Justice and Earned Income." *American Sociological Review* 42:639–651.

Jencks, Christopher. 1994. *The Homeless.* Cambridge, MA: Harvard University Press.

Jenson, Gary, and Maynard L. Erickson. 1979. "The Religious Factor and Delinquency: Another Look at the Hellfire Hypothesis." In *The Religious Dimension,* edited by Robert Wuthnow (pp. 157–177). New York: Academic Press.

Jindra, Michael. 1994. "Star Trek Fandom as a Religious Phenomenon." *Sociology of Religion* 55:27–51.

Johnson, Benton. 1963. "On Church and Sect." *American Sociological Review* 28:539–549.

Kalton, Graham. 1983. *Introduction to Survey Sampling.* Newbury Park, CA: Sage.

Kanter, Rosabeth Moss. 1972. *Commitment and Community.* Cambridge, MA: Harvard University Press.

Kiecolt, K. Jill. 1988. "Recent Developments in Attitudes and Social Structure." In *Annual Review of Sociology,* edited by W. R. Scott and Judith Blake. Palo Alto, CA: Annual Reviews.

Kifner, John. 1994. "Pollster Finds Error on Holocaust Doubts." *New York Times,* May 20, p. A6.

Kluckhohn, Clyde. 1949. *Mirror for Man.* New York: McGraw-Hill.

———. 1939. "On Recent Applications of Association Coefficients to Ethnological Data." *American Anthropologist* 41:345–377.

Kosmin, Barry A., Sidney Goldstein, Joseph Waksberg, Nava Lerer, Ariella Keysar, and Jeffrey Scheckner. 1991. *Highlights of the CJF 1990 National Jewish Population Survey.* New York: The Council of Jewish Federations.

Kosnick, Jon A. 1989. "Question Wording and Reports of Survey Results: The Case of Louis Harris and Associates and Aetna Life and Casualty." *Public Opinion Quarterly* 53:107–113.

Ko, Willem, Wim Meeus, and Harm t'Hart. 1991. "Religious Conversion of Adolescents: Testing the Lofland and Stark Model of Religious Conversion." *Sociological Analysis.* 52:227–240.

Langbein, Laura Irwin, and Allan J. Lichtman. 1978. *Ecological Inference.* Beverly Hills, CA: Sage.

LaPiere, Richard T. 1934. "Attitudes Versus Actions." *Social Forces* 12:230–237.

Laumann, Edward O., John H. Gagnon, Robert T. Michael, and Stuart Michaels. 1994. *The Social Organization of Sexuality: Sexual Practices in the United States.* Chicago: University of Chicago Press.

Lee, Raymond M. 1995. *Dangerous Fieldwork.* Thousand Oaks, CA: Sage.

Leedy, Paul D. 1993. *Practical Research: Planning and Design*. 5th ed. New York: Macmillan.

Lenski, Gerhard, and John Leggett. 1960. "Caste, Class, and Deference in the Research Interview." *American Journal of Sociology* 65:463–467.

Liebow, Elliot. 1967. *Tally's Corner*. Boston: Little Brown.

Linden, Rick, and Raymond F. Currie. 1977. "Religion and Drug Use: A Test of Social Control Theory." *Canadian Journal of Criminology and Corrections* 19:343–359.

Lofland, John. 1976. *Doing Social Life*. New York: Wiley.

———. 1966. *Doomsday Cult*. Englewood Cliffs, NJ: Prentice-Hall.

Lofland, John, and Lyn H. Lofland. 1984. *Analyzing Social Settings*. 2nd ed. Belmont, CA: Wadsworth.

Lofland, John, and Rodney Stark. 1965. "Becoming a World-Saver: A Theory of Conversion to a Deviant Perspective." *American Sociological Review* 30:862–875.

Lowenthal, Leo. 1944. "Biographies in Popular Magazines." In *Radio Research 1942–43*, edited by Paul F. Lazarsfeld and Frank N. Stanton (pp. 505–548). New York: Duell, Sloan and Pearce.

Lynxwiler, John, and David Gray. 1994. "Reconsidering Race Differences in Abortion Attitudes." *Social Science Quarterly*, 75:67–84.

MacKerras, Malcolm. 1977. "Do Women Candidates Lose Votes?" *Austrian Quarterly* 40:6–10.

Madamba, Anna B., and Gordon F. De Jong. 1994. "Determinants of White-Collar Employment: Puerto Rican Women in Metropolitan New York." *Social Science Quarterly*, 75:53–66

Malinowski, Bronislaw. 1948. *Magic, Science and Religion*. Garden City, NY: Doubleday.

Mandel, Michael J. 1994. "The Real Truth About the Economy: Are Government Statistics So Much Pulp Fiction?" *Business Week*, Nov. 7, pp. 110–118.

Meeker, Barabara Foley, and Robert K. Leik. 1995. "Experimentation in Sociological Social Psychology." In *Sociological Perspectives on Social Psychology*, edited by Karen S. Cook, Gary Alan Fine, and James S. House (pp. 630–649). Boston: Allyn and Bacon.

Meier, Paul. 1972. "The Biggest Public Health Experiment Ever." In *Statistics: A Guide to the Unknown*, edited by Judith M. Tanur. San Francisco: Holden-Day.

Melton, J. Gordon. 1981. "The Origins of Contemporary Neo-Paganism." Paper presented at the meetings of the Popular Culture Association, Detroit.

Menard, Scott. 1991. *Longitudinal Research*. Newbury Park, CA: Sage.

Merrill, John C. 1965. How *Time* Stereotyped Three U.S. Presidents." *Journalism Quarterly* 42:563–570.

Milgram, Stanley. 1974. *Obedience to Authority*. New York: Harper and Row.

Moore, Barrington, Jr. 1966. *The Social Origins of Dictatorship and Democracy*. Boston: Beacon.

Morrison, D. E., and R. E. Henkel, eds. 1970. *The Significance Test Controversy*. Chicago: Aldine.

Newman, Jody. 1994. *Perception and Reality: A Study Comparing the Success of Men and Women Candidates*. Washington, DC: National Women's Political Caucus.

O'Brien, Mary Utne, Sarah Segal Loevy, and Ann-Sofi Roden. 1986. *Data Collection Procedures*. Chicago: NORC.

Ostrom, Charles W. 1990. *Time Series Analysis*. Newbury Park, CA: Sage.

Patterson, Thomas E. 1994. *Out of Order*. New York: Vintage Books.

Ragin, Charles C. 1987. *The Comparative Method*. Berkeley: University of California Press.

Richer, Stephen. 1984. "Sexual Inequality in Children's Play." *Review of Canadian Sociology and Anthropology* 21:166–180.

Riesman, David, with Nathan Glazer and Reuel Denny. 1950. *The Lonely Crowd: A Study of the Changing American Character*. New Haven, CT: Yale University Press.

Robinson, William. 1950. "Ecological Correlations and the Behavior of Individuals." *American Sociological Review* 15:351–357.

Roethlisberger, Fritz Jules, and Willam John Dickson. 1939. *Management and the Worker*. Cambridge, MA: Harvard University Press.

Rosenthal, Robert, and Leonore Jacobson. 1968. *Pygmalion in the Classroom*. New York: Holt, Rinehart, and Winston.

Rossi, Peter H. 1989. *Down and Out in America*. Chicago: University of Chicago Press.

Rossi, Peter H., Richard Berk, and Kenneth J. Lenihan. 1980. *Money, Work, and Crime: Experimental Evidence*. New York: Academic Press.

Saenz, Rogelio, and Robert N. Anderson. 1994. "The Ecology of Chicano Interstate Net Migration, 1975–19980." *Social Science Quarterly*, 75:37–52.

Schatzman, Leonard, and Anselm L. Strauss. 1973. *Field Research: Strategies for a Natural Sociology*. Englewood Cliffs, NJ: Prentice-Hall.

Schele, Linda, and David Freidel. 1990. *A Forest of Kings: The Untold Story of the Ancient Maya*. New York: William Morrow.

Schollaert, Paul T., and Donald Hugh Smith. 1987. "Team Racial Composition and Sports Attendance. *Sociological Quarterly* 28:71–87.

Schuman, Howard, and Stanley Presser. 1981. *Questions and Answers in Attitude Surveys: Experiments on Question Form, Wording, and Context*. New York: Academic Press.

Selltiz, Claire, Marie Johoda, Morton Deutsch, and Stuart W. Cook. 1959. *Research Methods in Social Relations*. New York: Holt, Rinehart, & Winston.

Shaffir, William B., Robert A. Stebbins, and Allan Turowetz, eds. 1980. *Fieldwork Experiences: Qualitative Approaches to Social Research*. New York: St. Martin's Press.

Sherman, Lawrence W., and Richard A. Berk. 1984. "The Specific Deterrent Effects of Arrest for Domestic Violence." *American Sociological Review* 49:261–272.

Shorter, Edward. 1975. *The Making of the Modern Family*. New York: Basic Books.

Skocpol, Theda. 1979. *States and Social Revolutions: A Comparative Analysis of France, Russia and China*. Cambridge: Cambridge University Press.

Skolnick, Jerome. 1966. *Justice Without Trial: Law Enforcement in Democratic Society*. New York: Wiley.

Smelser, Neil J. 1976. *Comparative Methods in the Social Sciences*. Englewood Cliffs, NJ: Prentice-Hall.

Smith, Jane S. 1990. *Patenting the Sun: Polio and the Salk Vaccine*. New York: Morrow.

Smith, Tom W. 1983. "The Hidden 25 Percent: An Analysis of Non-Response to the 1980 General Social Survey." *Public Opinion Quarterly* 47:386–404.

———. 1979. "Sex and the GSS." *General Social Survey Technical Report #17*. Chicago: NORC.

Sorenson, Elaine, and Frank D. Bean. 1994. "The Immigration Reform and Control Act and the Wages of Mexican Origin Workers: Evidence from Current Population Surveys." *Social Science Quarterly* 75:1–17.

Stark, Rodney. 1996. "Religion as Context: Hellfire and Delinquency One More Time." *Sociology of Religion* 57:163–173.

———. 1991. "Christianizing the Urban Empire: An Analysis Based on 22 Greco-Roman Cities." *Sociological Analysis* 52:77–88.

———. 1987. "Deviant Places: A Theory of the Ecology of Crime." *Criminology* 25:891–907.

Stark, Rodney, and William Sims Bainbridge. 1985. *The Future of Religion: Secularization, Revival, and Cult Formation*. Berkeley: University of California Press.

Stark, Rodney, and Charles Y. Glock. 1965. "The 'New Denominationalism.'" *Review of Religious Research* 7:8–17.

Stark, Rodney, Lori Kent, and Daniel P. Doyle. 1982. "Religion and Delinquency: The Ecology of a 'Lost' Relationship." *Journal of Research in Crime and Delinquency* 19:4–24.

Statistics Canada. 1991. *User's Guide to the General Social Survey, "Family and Friends," 1990*. Ottawa.

Stouffer, Samuel A., Louis Guttman, Edward A. Suchman, Paul F. Lazarsfeld, Shirley A. Star, and John A. Clausen. 1950. *Measurement and Prediction*. Princeton, NJ: Princeton University Press.

Stouffer, Samuel A., Arthur A. Lumsdaine, M. H. Lumsdaine, Robin M. Williams, Jr., M. Brewster Smith, Irving L. Janis, Shirley A. Star, and Leonard S. Cottrell, Jr. 1949. *The American Soldier: Combat and Its Aftermath, vol. II.* Princeton, NJ: Princeton University Press.

Stouffer, Samuel A., Edward A. Suchman, L. C. Devinney, Shirley A. Star, and Robin M. Williams, Jr. 1949. *The American Soldier: Adjustment During Army Life, vol. I.* Princeton, NJ: Princeton University Press.

Strunk, William, Jr., and E. B. White. 2000. *The Elements of Style.* 4th ed. New York: Allyn and Bacon.

Taeuber, Karl E. 1983. *Report of the Citizens' Cimmission on Civil Rights.* Washington, DC: Center for National Policy Review, Catholic University.

Tilly, Charles. 1984. *Big Structures, Large Processes, Huge Comparisons.* New York: Russell Sage Foundation.

———. 1969. "The Analysis of a Counter-Revolution." In *Quantitative History,* edited by Don Karl Rowney and James Q. Graham, Jr. (pp.181–208). Homewood, IL: The Dorsey Press.

Tilly, Charles, and Louise Tilly. 1974. *The Rebellious Century.* Cambridge, MA: Harvard University Press.

Tolnay, Stewart E., and E. M. Beck. 1992. "Racial Violence and Black Migration in the American South, 1910 to 1930." *American Sociological Review* 57:103–116.

Tourangeau, Roger, Robert A. Johnson, Jiahe Qian, and Hee-Choon Shin. 1993. *Selection of NORC's 1990 National Sample.* Chicago: NORC.

Trent, Katherine, and Scott J. South. 1989. "Structural Determinants of the Divorce Rate: A Cross-Cultural Analysis." *Journal of Marriage and the Family* 51:391–404.

Tuckel, Peter S., and Barry M. Feinberg. 1991. "The Answering Machine Poses Many Questions for Telephone Survey Researchers." *Public Opinion Quarterly* 55:200–217.

U.S. Department of Health, Education, and Welfare. 1960. *Report to Congress on Juvenile Delinquency.* Washington, DC: U.S. Government Printing Office.

Wallerstein, Immanuel. 1974. *The Modern World System.* New York: Academic Press.

Webb, Eugene J., Donald T. Campbell, Richard D. Schwartz, and Lee Sechrest. 1981. *Unobtrusive Measures: Nonreactive Research in the Social Sciences.* Boston: Houghton Mifflin.

Weeks, Michael F., and R. Paul Moore. 1981. "Ethnicity of Interviewer Effects on Ethnic Respondents." *Public Opinion Quarterly* 45:245–249.

White, Lynn K., and John N. Edwards. 1990. "Emptying the Nest and Parental Well-Being: An Analysis of National Panel Data." *American Sociological Review* 55:235–242.

Whyte, William Foot. 1943. *Street Corner Society.* Chicago: University of Chicago Press.

Wilson, Thomas C. 1991. "Urbanism, Migration, and Tolerance: A Reassessment." *American Sociological Review* 56:117–123.

Yammarino, Francis J., Steven J. Skinner, and Terry L. Childers. 1991. "Understanding Mail Survey Response Behavior: A Meta-Analysis." *Public Opinion Quarterly* 55.613–639.

Glossary

abstraction An intellectual creation, a definition, existing only in our minds. Because scientific concepts are abstract they apply to all possible members of the class, all that have been, are, shall be, or could be.

acquiescence With reference to survey research, acquiescence is the tendency of some respondents to say "Yes" or "Agree" (some experts refer to this as yeah-saying).

age effects Effects that occur because people change as they get older. For example, people may become more satisfied with their lives as they mature.

aggregate data To aggregate is to put together; hence, aggregate data refer to data based on units of analysis larger than the individual. Because, in a sense, aggregate data describe the social reality surrounding the individual, such data also often are referred to as ecological data.

alternate forms The approach that assesses reliability by comparing two or more independent measures of the same variable.

antecedent variable The cause of spurious correlations between other variables. The word *antecedent* is defined as *going before, prior, or preceding*. We call the source of spurious relationships *antecedent variables* because, as causes, these variables must come before their consequences.

applied research The primary purpose of applied research is to serve practical needs. It may or may not involve theory testing but most often does not.

areal units Areal units consist of aggregate units having geographic boundaries—an area.

attitude An enduring, learned predisposition to respond in a consistent way to a particular stimulus or set of stimuli. We all have attitudes toward significant things in our environment: objects, people, ideas, events, and so on. An attitude usually is regarded as consisting of three components: (1) a belief, (2) a favorable or unfavorable evaluation, and (3) a behavioral disposition (DeLamater, 1992).

behavior Behavior involves action; it is what living creatures *do*. We "hold" attitudes, but we "perform" behavior.

birth cohorts People born within the same period of time.

cafeteria question A question that allows respondents to pick and choose from a large selection of responses, indicating only the ones they select.

case-oriented approach This approach selects two or more (but rarely more than five) cases and examines them closely in order to explain some striking difference or differences between (or among) them.

categorical variables Variables that sort cases into categories. Gender is a categorical variable; cases can be separated into one of two categories: male or female. However, the categories of a nominal variable lack intrinsic *order*. Also called *nominal* variables.

causal models Statistical descriptions of a specific set of empirical data that attempt to identify and measure all of the relationships among some set of independent variables and the dependent variable. Often, a causal model is designed to gauge dynamic relationships among the variables based on deductions from a theory.

cause Anything producing a result or an effect, as when one variable produces or results in variation in another variable.

census As defined in dictionaries, a census is an official count of the population and the recording of certain information about each person. In social science, the term is used more broadly and refers to instances when *data are collected from all cases or units in the relevant set.*

clear boundaries Delineations that eliminate ambiguity about what a concept does and does not include efficient concepts will have clear boundaries.

closed questions Questions that force all respondents to select their responses from a set the researchers provide.

cluster sampling A two-step process in which aggregates or groups of individuals (clusters) are sampled, and then the samples of individuals are selected from within each aggregate (cluster).

cohort effects Older people may differ from younger people, not because they changed as they got older, but because their generation always differed from later generations. These aspects of change are called cohort effects.

cohorts Persons within some subgroup of a population who share a significant life experience or event within a given period of time, usually from one to ten years. Usually, the shared experience is birth, and nearly all cohort analysis in social science is based on birth cohorts—people born within the same period of time.

comparative research While all research is based on comparisons, the term comparative research usually identifies studies based on aggregate units of analysis—nations, states, cities, counties.

concepts Scientific concepts are abstract terms that identify a class of "things" to be regarded as alike.

confidence interval The *range* within which we estimate the statistic to depart from the parameter.

confidence level The *probability* that the parameter actually will fall within the range stated by the confidence interval.

conformity As it pertains to survey research, conformity refers to the tendency of respondents to select answers on the basis of their perceived social standards, to select "normal" or noncontroversial answers.

constants Characteristics or aspects of the things being studied that do not vary, but that take the same value. All cities have crimes; therefore, having crimes is a constant, just as breathing is a characteristic of all living human beings.

construct validity Construct validity is based on the match-up between a measure and the assumptions concerning the phenomena the concept is meant to isolate and identify. For example, the Herfindahl index used to measure religious pluralism rests on an elaborate mathematical model from which its computational formula is derived. In such cases, it often is possible to test the validity of a measure by seeing if the measure meets these underlying assumptions.

content analysis A research technique used to systematically transform nonquantified verbal, visual, or textual material into quantitative data to which standard statistical analysis techniques may be applied. For example, content analysis has been applied to books, diaries, speeches, stage plays, newspaper and magazine articles, song lyrics, television and radio programs, movies, ads, and even graffiti.

contingency questions Questions that divert respondents to different questions or direct them to skip questions, *contingent* on responses to a prior question.

control group A group that consists of those persons randomly assigned to not be exposed to the experimental stimulus. It's purpose is to provide a baseline against which to assess the effect of the experimental stimulus.

convergent validity A test of validity based on the principle that valid measures of the same concept must be correlated. This is called convergent validity because the indicators converge on a single, underlying, empirical base that represents the concept.

correlated If something is the cause of something else, then it *must* be the case that the cause and effect *vary in unison*. Changes in the cause must produce changes in the proposed effect. When variables vary or change in unison, they are correlated.

correlation Correlation means "to go together, to vary in unison." Correlations can be either positive or negative. If sales of ice cream rise when the temperature rises and fall when the temperature falls, there is a positive correlation. However, if the sales of down jackets fall when the temperature gets warmer and rise when it gets colder, there is a negative correlation.

covert observation When those being observed are unaware that they are the objects of research, researchers are engaging in covert observation.

criterion validity The most stringent test of validity. It is based on comparing a particular indicator with another that is *known to be valid* and which can, therefore, serve as a *criterion*.

Cronbach's alpha The average inter-item correlation for a set of items.

cross-case comparability Cross-case comparability refers to the need that a variable measure the same thing for each case or unit of analysis.

cross-sectional study A study that represents one point in time.

crude birth rate The number of births in a given year per 1,000 population.

cultural artifact An artifact is any object made by human work, and a cultural artifact is any such object that informs us about the physical and/or mental life of some set of human beings. The word *object* is interpreted very broadly to include written, filmed, and recorded material.

data In science, observations are referred to as data whereas a single observation is a datum.

deduction Deduction involves reasoning from the general to the specific—from a known (or assumed) principle to an unknown but observable conclusion. Put another way, when scientists use deductive logic they begin with a general, abstract premise or proposition and then show that less general, empirical predictions are implied by the statement according to the rules of logic.

demand characteristic In experiments, a demand characteristic is an aspect of the design that tips off subjects as to what they are expected to do. Subjects tend to be cooperative and, therefore, to respond to demand characteristics.

dependent variable A dependent variable is hypothesized to be the effect *being caused*. That is, variations in delinquency are hypothesized to be caused by variations in family structures—variations in delinquency are *dependent* on variations in family structure.

deposit bias Deposit bias refers to circumstances in which only some portion of the pertinent material was originally *included* in the set of materials to be coded.

dichotomy Anything that consists of only two values.

disproportionate stratification See oversample.

double-blind experiment An experiment in which neither the subjects nor the experimenters know whether or when the independent variable varies.

dummy variable A categorical variable recoded to assign the values 0 and 1 for *each category* so that 1 indicates the presence of the category and 0 represents its absence. When categorical variables have more than two categories, then more than one dummy variable must be created. However, when creating dummy variables, *never* create more than the total number of categories *minus one.*

ecological data See aggregate data.

ecological fallacy When we assume that findings based on aggregate or ecological data apply to individuals, we are committing the ecological fallacy.

empirical Empirical means "observable through the senses."

empirical generalization A summary statement based on empirical observations. "Men are more likely than women to favor capital punishment" is an empirical generalization summing up the results of many surveys.

experimental group A group that consists of those persons randomly assigned to be exposed to the experimental stimulus—to the independent variable.

experiments Experiments have two fundamental features: (1) the researchers are able to *manipulate* the independent variable (making it vary as much as they wish, whenever they wish), and (2) there is random assignment of persons to groups exposed to different levels of the independent variable. People who take part in an experiment often are referred to as the **subjects** because they are subjected to different values of the independent variable.

exploratory research Research that occurs when social scientists make systematic observations of uncharted and little-known phenomena in order to get an initial sense of what is going on.

external validity The extent to which the findings of an experiment can be generalized.

extreme outlier A case having such a deviant value on a variable that it distorts correlations, sometimes causing an apparent correlation where none exists among the other cases and sometimes suppressing a correlation that does exist among the other cases.

evaluation research Research that is conducted to assess the effectiveness of a program, policy, product, or procedure. It usually is commissioned by government agencies, businesses, or organizations such as schools, churches, or hospitals. If the program, policy, procedure, or product being evaluated is based on a theory, then evaluation studies also involve theory testing. But this usually is not the case.

face validity The most common basis for establishing validity—to conclude that *on the face of it* the variable obviously measures the concept.

fallacy of unmanipulated causes The fallacy that involves mistaking the effects of individual *characteristics of subjects* on the dependent variable for experimental effects.

falsify To collect and analyze data that show that a hypothesis is incorrect (or false).

fertility rate The number of births to the average female during her lifetime (this is estimated in a variety of ways).

field notes Written accounts of what a field researcher sees and hears. They should be made as soon as possible after the observations and should be as complete as possible.

field research Research that involves going out to observe people as they engage in the activities the social scientist wants to understand. This research is called *field research* because it is conducted in the field—in the natural settings in which the people and activities of interest are normally to be found. Sometimes, field research is guided by hypotheses, but often it is exploratory and produces suggestive hypotheses only after the observations are completed.

forced option question A question that requires a response to every option.

history In the context of an experiment, history refers to any event occurring simultaneously with the independent variable that might account for the observed changes.

hypothesis This tells us what to expect to observe when we examine relationships among indicators. If the hypothesis is derived from a theory, the theory specifies relationships among concepts and this is reflected in the hypothesis, which specifies the relationships to be observed among indicators. Even when hypotheses derive from hunches or common sense rather than from theories, they specify where we should look and what we ought to observe.

independent variable An independent variable is hypothesized to be the *cause* of something else. Thus, variation in family structures is hypothesized to cause variations in delinquency scores.

indexes Combinations of variables based on survey data often are referred to as indexes (or indices). When social scientists create indexes based on individual level data (such as survey questions), they typically proceed simply by adding together the values of some set of indicators. Such an index is based on the assumption that *more* measures of the same thing will yield more sensitive measurements.

indicator An observable measure of a concept.

induction Involves reasoning from the specific to the general, from a set of observations to a general conclusion. Empirical generalizations are obtained through induction.

informants Persons assumed to be well informed on matters of interest to a field researcher and who, therefore, are asked to provide information about a group as a whole or about particular members or subgroups.

Instrument effects Differences produced by variations in the accuracy of multiple measurements.

internal consistency This method assesses the reliability of a set of measures of the same concept by comparing all possible combinations of these items and calculating Cronbach's alpha, which is the average inter-item correlation for a set of items.

internal validity Internal validity consists of eliminating all of the sources of bias. That means there must be at least two groups to compare (two levels of the independent variable including present and absent) and that assignment to groups must be random.

inter-rater reliability Inter-rater reliability assumes that a variable is well measured if it is created or scored by independent raters or coders who achieve a high level of agreement.

interval variable A more precise form of ordinal variable in that the gaps or intervals between categories are of equal quantity.

intervening variable An intervening variable is hypothesized to be the *link* between an independent and a dependent variable.

lagged time series Whenever the values of the independent variables or the dependent variables assigned to specific cases come from different time periods—when there is a systematic time lag among variables—this is a lagged time series.

latent content Refers to "deeper" or *implicit* meanings.

law of parsimony As used by modern philosophers of science, the law of parsimony (sometimes referred to as Ockham's razor) reads: *Theories always should attempt to explain the most with the least*. We should try to explain as much as possible with a theory that is as simple as possible. Applied to concepts, parsimony encourages greater abstraction.

longitudinal studies Longitudinal or panel studies are based on surveys of the *same respondents* over a period of time, sometimes 10 years or more, to see how they have changed. Panel studies usually span a rather short time, while longitudinal studies tend to cover relatively longer periods.

manifest content The *explicit*, clear, and perhaps superficial meaning of verbal, visual, or textual materials.

maturation In experiments, maturation refers to the effects of subjects getting older (maturing) as time passes.

necessary cause When an effect *never* occurs in the absence of a particular independent variable, we refer to this as a necessary cause or condition.

nominal definitions As Carl G. Hempel (1952) pointed out in his classic work on scientific concepts, it sometimes is claimed that "real" definitions somehow are an intrinsic part of the thing being defined, but this is folklore. Scientific definitions are not thought of as real in this sense at all, but are regarded as names that simply are *assigned* to something. To underscore this point, Hempel identified all scientific concepts as nominal definitions—that they are merely names.

nominal variables Variables that sort cases into categories. Gender is a nominal variable; cases can be separated into one of two categories: male or female. However, the categories of a nominal variable lack intrinsic *order*. Also called *categorical* variables.

nonlagged time series When the independent and dependent variables are measured at the same point in time for any given case, this is a nonlagged time series.

null hypothesis A null hypothesis asserts that a hypothesis derived from a theory is false. That is, lacking persuasive evidence to the contrary, we accept the null hypothesis which states that there is no correlation among indicators of a theory.

Ockham's razor See *law of parsimony*.

open-ended questions Questions that permit respondents to answer as they wish, in their own words.

operationalizing The process of selecting indicators of concepts.

ordinal variables Variables whose categories can be ordered. Thus, for example, people can be ranked along a continuum from liking jazz very much to disliking it very much.

oversample Researchers often oversample the smaller strata; that is, they select more cases from a small stratum than its true population proportion—a practice sometimes referred to as disproportionate stratification.

overt observation When the identity of the observers as social researchers is known to those being observed, researchers are using overt observation.

panel studies Panel or longitudinal studies are based on surveys of the *same respondents* over a period of time, sometimes 10 years or more, to see how they have changed. Panel studies usually span a rather short time, while longitudinal studies tend to cover relatively longer periods.

parameter A parameter refers to the *true value* of a variable within the population or universe. For example, the actual mean income of a population might be $23,789 per year. A **statistic** refers to the *observed value* of a variable within the sample. Thus, based on a sample, we might estimate the average income of a population to be $23,186 per year. In this example, the statistic slightly underestimates the population parameter, or the true value.

period effects Surveys may find substantial differences over time because *everyone* is changing. These are called period effects to indicate that they reflect the influence associated with a particular historical period.

placebo As used in experiments, a placebo is a substitute for the independent variable and is used to permit variation in the independent variable without permitting variation in other factors linked to the independent variable. For example, in an experiment to test a drug, one treatment group gets the drug while the other gets a harmless substitute, but *both* groups get injected.

population As used by social scientists, the word *population* is not limited to human beings, but consists of *all units* constituting a set, however that set is defined or delimited. A population also sometimes is referred to as the **universe** of units, in that the word *universe* refers to "all things." "All persons in Denmark" defines a population or universe as does "all children in the fifth grade at Washington School" or "all counties in the United States."

probability proportional to size (PPS) Compensate for clusters having different numbers of members and involves selecting a sample from each cluster based on the relative size of each cluster.

proportional fact A proportional fact asserts the distribution of something, or even the joint distribution of several things, among a number of cases.

pure research Pure research (or basic research) is directed by the desire to increase knowledge, without regard for potential practical applications. Such research usually involves theory testing and/or efforts to refine or extend theories.

quasi-experiments Studies posing as experiments, but which fail to meet the essential experimental standards of internal validity. (*Quasi* means "having superficial resemblance to.")

R^2 Measures the *combined*, or joint, effects of the two independent variables on the dependent variable. R^2 can be converted to a percentage by moving the decimal point two places to the right.

random selection All cases have an equal chance, or at least a known probability, of being included in a sample.

rate A proportion or ratio that usually is created by dividing one variable by another variable. The purpose of rates is to create a common basis for comparison across aggregate units so as to allow for differences in such factors as size.

ratio variables Variables that have meaningful zero points. Thus, we can say that a person weighing 220 pounds is twice as heavy as a person weighing 110 or that someone age 40 is twice as old as someone age 20.

regression The statistical technique that uses the correlations between each pair of variables in a set of variables to calculate relationships among the entire set.

regression to the mean This refers to the tendency of extreme scores to move (or regress) over time toward the group average (or mean).

reliable A variable is reliable if it is consistent—if repeated observations give similar results.

replication research The repetition of previous research to check on the results.

response set The tendency of respondents to fall into a pattern of responding, having no regard for variations in content.

sample A set of units or cases *randomly selected* from a population or universe. *Random selection* means that all cases have an equal chance, or at least a known probability, of being included in the sample. The fundamental principle on which sampling rests is: *If all cases have a known probability of being selected for inclusion in a sample, then we can calculate the probability that the group included in the sample is identical to (or representative of) those not included.* Random samples "work" because they are based on the laws of probability, and, therefore, social researchers tend to use the terms *random sample* and *probability sample* interchangeably.

scales Measures created by combining variables using an explicit measurement model. Such models typically include three elements:

> (1) assumptions about the nature of the component questions, or variables, (2) rules for combining the set of questions, or variables, to produce the measure, and (3) techniques for assessing the quality of the resulting measure, or scale.

selection biases Biases that occur whenever people are assigned to the experimental or control groups (or to groups based on levels of the independent variable) in nonrandom ways.

simple fact An assertion about a concrete and limited state of affairs, often merely the claim that something happened or exists and usually having to do with only one or very few cases.

simple random sampling A technique based on the principle that all members of the population have an equal chance of being selected.

SLOPS Self-selected listener opinion polls.

snowball sample A "sample" that is assembled by referral, as persons having the characteristic(s) of interest identify others.

social units Social units consist of aggregate units having social boundaries—the basis of inclusion is social, not geographic. All persons residing within a clearly designated area belong to the aggregate making up Chicago. But, only a few of these people belong to the aggregates making up the Chicago White Sox or Cubs.

split halves The technique that separates measures collected simultaneously and compares them.

spurious Two variables often appear to have a cause-and-effect relationship when, in fact, they are correlated *only* because each is correlated with some third, unobserved or unnoticed variable. Correlations such as this are called spurious. They appear to reflect causation, but they don't.

standardized beta The standardized beta estimates *the independent effect of each independent variable on the dependent variable*. The value of beta is, therefore, the *net* effect of each independent variable on the dependent variable.

statistic The observed value of a variable within the sample.

statistical generalization Statistical generalization involves successfully inferring that observations based on a sample apply to the unobserved members of that same population.

stratified random samples Social researchers often divide a population into several subpopulations or strata, based on information about each unit or case, and then select samples independently from each. This technique is referred to as selecting stratified random samples.

stratified sampling by characteristics Stratified sampling by characteristics requires that we know the actual proportion of each stratum in the population and that it be possible to draw separate samples from each stratum.

structured observations Focused and intentional observations that often are recorded on forms prepared for that purpose.

subject "mortality" This refers to the loss of subjects over the duration of an experiment.

subjects People who take part in an experiment often are referred to as the subjects because they are subjected to different values of the independent variable.

sufficient cause When an effect *always* occurs when a particular independent variable is present, we refer to this as a sufficient cause or condition.

suppressor variable A variable that causes two variables to appear *not to be correlated* when in fact they are. In a sense this is the opposite of spuriousness.

survey research Research that is based on samples of individuals who are interviewed or who fill out questionnaires. People included in the survey often are referred to as respondents since they respond to the questions asked by the interviewer or included in the questionnaire.

survival bias Survival bias refers to circumstances in which only some portion of the pertinent phenomena was *retained* in the set of materials to be coded.

systematic random sampling A variant on random sampling. It involves sampling a list by selecting the first case randomly and then taking every *n*th case until the end of the list is reached. First, the researcher must number a list of the entire population and then divide the total number in the population by the number desired for the sample, thus obtaining a *sampling fraction*.

tautology Any statement that is true by definition as in President Calvin Coolidge's famous remark that "as more and more men are thrown out of work, unemployment results."

testing effects Effects that occur when the tests or measurements are themselves causative agents—whether by changing views or by influencing performance, as in the case of subjects becoming testwise.

test of significance A calculation of the odds that a difference or correlation is produced by random fluctuations between the sample and the population, between the parameter and the statistic.

test-retest reliability Test-retest reliability involves measuring the same cases at two or more times and comparing the results.

theoretical generalization Theoretical generalization involves increasing the scope (generality) of tests of a theory by applying the theory to a variety of settings.

theories Abstract statements saying *why* and *how* some set of concepts are linked. Their purpose is to *explain* some portion of reality. Because theories contain concepts and concepts are abstractions, all scientific theories also are abstractions. That is, you can't see or touch a theory. There is one more vital feature of real theories: They must make empirical predictions and prohibitions—that is, it must be possible to say what sorts of observations would *falsify* the theory.

time order The sequence of variables. As a criterion of causation, it involves a very simple principle: A cause must occur *before* its effect. Put another way, the principle of cause-and-effect makes no sense backwards.

time series analysis Time series analysis examines the relationship between two or more variables based on the same universe but measured at a number of points in time. Therefore, in time series analysis, *points in time* are the *units* of analysis.

trend studies Studies that are based on two or more surveys conducted at different times, but based on independently selected samples of the *same population* and which include identical (or adequately comparable) questions.

type one error Type one error involves accepting that the predicted correlation exists, when in fact it does not.

type two error Type two error involves concluding that the predicted correlation does not exist, when in fact it does.

units of analysis Social scientists base their research on a variety of units of analysis—the "things" a hypothesis directs us to observe. Often these things are individual human beings which are referred to as *individual* units. But, they also consist of larger units which are referred to as *aggregate* units.

unobtrusive measure A measure that has no effect on the objects being studied—it is a measurement that does not intrude. Such measures sometimes are referred to as *nonreactive measures*—ones that produce no reaction.

unstructured observations Informal, often impromptu observations that usually are recorded in a narrative fashion.

utility The ultimate test of all concepts. Utility means nothing more nor less than whether or not they turn out to be useful in constructing efficient theories. *What works better, is better!*

valid A variable is valid if it actually measures the concept it is meant to measure.

variables Characteristics or aspects that take different values among the things being studied and thus can be said to vary. Homicide rates vary across the 50 states, and therefore the homicide rate is a variable as are all aspects of states on which they differ including the proportions of their populations who bowl, eat pizza, or drive pickup trucks. At the individual level, all ways in which people differ are variables: height, weight, religion, hobbies, voting preferences, and the like.

variable-oriented approach The approach that tests hypotheses by applying statistical techniques such as correlation and regression to variables based on an appropriate set of aggregate cases.

wave Each instance of data collection from respondents in a panel or longitudinal survey.

weighting The assigning of different values (or weights) to each case in order to restore proper proportionality to the sample.

Index